Studies in Empirical Economics

Aman Ullah (Ed.)
Semiparametric and Nonparametric Econometrics
1989. VII, 172 pp. Hard cover DM 120,-
ISBN 3-7908-0418-5

Walter Krämer (Ed.)
Econometrics of Structural Change
1989. X, 128 pp. Hard cover DM 85,-
ISBN 3-7908-0432-0

Wolfgang Franz (Ed.)
Hysteresis Effects in Economic Models
1990. VIII, 121 pp. Hard cover DM 90,-
ISBN 3-7908-0482-7

John Piggott and John Whalley (Eds.)
Applied General Equilibrium
1991. VI, 153 pp. Hard cover DM 98,-
ISBN 3-7908-0530-0

Baldev Raj and Badi H. Baltagi (Eds.)
Panel Data Analysis
1992. VIII, 220 pp. Hard cover DM 128,-
ISBN 3-7908-0593-9

Josef Christl
The Unemployment / Vacancy Curve
1992. XVI, 152 pp. Hard cover DM 98,-
ISBN 3-7908-0625-0

Jürgen Kaehler and Peter Kugler (Eds.)
Econometric Analysis of Financial Markets
1994. VII, 230 pp. Hard cover DM 98,-
ISBN 3-7908-0740-0

Klaus F. Zimmermann (Ed.)
Output and Employment Fluctuations
1994. VI, 259 pp. Hard cover DM 128.-
ISBN 3-7908-0754-0

Jean-Marie Dufour and Baldev Raj (Eds.)
New Developments in Time Series Econometrics
1994. V, 250 pp. Hard cover DM 148,-
ISBN 3-7908-0766-4

John D. Hey (Ed.)
Experimental Economics
1994. VII, 136 pp. Hard cover DM 98,-
ISBN 3-7908-0810-5

Arno Riedl, Georg Winckler and Andreas Wörgötter (Eds.)
Macroeconomic Policy Games
1995. XII/123 pp. Hard cover DM 90,-
ISBN 3-7908-0857-1

Thomas Url
Andreas Wörgötter (Eds.)

Econometrics of Short and Unreliable Time Series

With 78 Figures

Physica-Verlag
A Springer-Verlag Company

Editors
Dr. Thomas Url
Austrian Institute of Economic Research
P.O. Box 91
A-1103 Vienna, Austria

Univ.-Doz. Dr. Andreas Wörgötter
Head of Department of Economics
Institute for Advanced Studies
Stumpergasse 56
A-1060 Vienna, Austria

ISBN-13: 978-3-642-99784-6 e-ISBN-13: 978-3-642-99782-2
DOI: 10.1007/978-3-642-99782-2

CIP-Titelaufnahme der Deutschen Bibliothek
Econometrics of short and unreliable time series / Thomas Url
; Andreas Wörgötter (ed.). - Heidelberg: Physica-Verl., 1995
(Studies in empirical economics)

NE: Url, Thomas [Hrsg.]

Softcover reprint of the hardcover 1st edition 1995

88/2202-5 4 3 2 1 0 - Printed on acid-free paper

Contents

Introduction and Overview

Thomas Url

Motivation

Short and unreliable time series seem to be an odd starting point for any statistical inference. Additional observations and regularity of the underlying economic structure are usually supposed to be a prerequisite of any meaningful empirical analysis. In such a case the available data can be presumed to come from a random experiment and adding observations increases our experience and reduces the sampling error. Although an exact experimental design will never be available in empirical economics, it should be possible to formulate stable causal relationships among variables. In this sense any statistical model is composed of a joint probability density function with a known mathematical form and a known parameter space. Statistical inference then consists of using the sample values to estimate plausible parameter values. Therefore, empirical applications are usually based on a minimum number of observations with more sophisticated methods often requiring a growing number of observations.

A study of the econometrics of short and unreliable time series is especially interesting in light of the emergence of a large group of countries on the world market, in which economic information is only available for quite a short period. Nevertheless, market participants from those economies and from developed countries as well as governmental bodies would like to base their economic decisions on a proper picture of their aggregate economic behavior. Examples that come immediately to mind are the countries in transition from planned economies to market economies. Located either in Eastern Europe, in the former Asian part of the Soviet Union or in South East Asia, those economies suffer from two substantial problems with regard to any application of statistical or econometrical methods.

First of all, the accuracy of measurement is probably not as high as could be expected from developed economies with established statistical agencies. One of the main sources of difficulties is that planned economies developed completely different data demands. Planning activities required very detailed information on individual firms and production processes. Therefore the popularity of Input-Output analysis in planned economies. Moreover, the economic structure was adapted to planning requirements and assumed increasing returns to scale technologies. Firms were provided with targets to be met and received a specified amount of resources to fulfill their task. In this environment there is no need for any further information on distribution networks, consumer tastes and reactions, price responsiveness and the like. Aside from these different needs, political interests prevented a documentation of price developments and financial data. On the other hand, certain problems were actually of no interest. For example, these countries published no labor market statistics at all, because full employment was guaranteed by the government and firms did not bother with efficiency issues due to lack of or non-enforced budget constraints.

The second problem concerns the rapid economic development in these countries during their transition towards a market economy. In such an environment the data generation process undergoes substantial changes, implying variations in the parameters of the structural form and simultaneously unstable parameters in the reduced form of a statistical model. For example, the break down of the rouble based foreign trade destroyed former trading relations and the official foreign trade institutions with one blow. Moreover, from official publications it is generally not known to what extent the private sector is correctly reflected in the data or whether there is a huge black market component in output. The most important changes come from the shift in ownership structures by privatization, joint ventures, and the creation of new private enterprises. This will change behavioral attitudes as well as the applied technologies.

The statistical problems are aggravated by the actual state of economic theory, which does not yet provide complete models for transition economies. With these fundamental requirements violated, one is tempted to avoid any statistical inference and to rely completely on personal and judgmental assessments. Yet it is exactly the vagueness and the subjectivity associated with such an approach that makes other methods based on statistics or econometrics interesting. Every analyst will come up with different conclusions from a given set of information. Statistical methods aim at standardizing the way in which a given set of information is analyzed and thus should provide a means for communicating the results of personal reasoning. Of course there is a trade off between the clarity of an analysis and its flexibility and the user has to decide which approach to favor.

This study consists of both theoretical and empirical contributions on short and unreliable time series. Various ways to solve those problems are presented by authors from Eastern Europe and from Western free market economies. This will give the reader an impression of the applicability of statistical methods as well as suggestions for problem solutions and first hand empirical results. After a short overview on the availability and quality of data from Eastern European countries the volume proceeds with the possibilities of achieving reasonable results within standard macroeconometric modeling techniques. Another string of ideas encompasses time series models with flexible trend methods. Such models should be able to adapt quickly to newly emerging situations. Data with higher periodicity are hard to gather and often only comprise a period of slightly more than two or three years. The possibilities and consequences of changing the periodicity of available data are also considered within this part. Empirical applications to Eastern European data are presented in several contributions. Two papers specializing on this issue form the last part of this volume.

I. The Econometric Approach: An Overview

The first section starts with the data problems in Eastern European countries and proposes an improvement of statistical analysis by a pretreatment of the data. A classical example is provided by Andrei Kazmin who analyzes monetary indicators in the Soviet Union with regard to the drastic shortage of reliable published data. He stresses the fact that official statistics have been used to foster group interests and to hide price increases. Moreover, official statements often tried to establish the image of a stable financial system by manipulating statistics. Kazmin then provides some information on how to correct wrong information and duplicate counting in published data and how to collect information on data yet to be published.

In her paper on macroeconomic forecasting, Judit Neményi begins with the assertion that targeting economic activity contradicts the fundamental idea of econometric market analysis because the data do not reflect the reaction of independent agents to price signals within an atomistic free market setting. Although there are no general a priori guidelines on how to cope with these difficulties, she presents the models used at the Hungarian Economic Research Institute and gives an idea of how to combine behavioral equations for supply and demand with the exogeneity assumption for prices. By using business survey data and other information on tensions in capacity utilization, the price formation process can be integrated into the econometric model.

The starting point of Edward Leamer's contribution is the limitation of available data sets in terms of quantity, quality and relevance. Faced with these scarcities a user may import data from analogous Western countries. Leamer suggests a Bayesian approach of pooling Western and Eastern data in a linear regression model. The main links between Western and Eastern countries are the degree of confidence in prior information, the degree of similarity of economic structures, and the amount of experimental contamination by poor quality or irrelevant data, omitted variables or other pathologies. Such a strategy will facilitate the conversation about the strength and the importance of analogies on which the pooling depends. A sensitivity analysis at the end of his paper highlights the impact of variations in prior parameters on estimated regression coefficients.

Another approach to pooling data from Eastern Europe with the economic structure from a Western European economy is presented by Jan Veverka. In Czechoslovakia users can build on a solid data base for wages and prices but the old models of the price determination process concentrated on the purpose of providing good cost estimates at highly disaggregated levels. Because prices used to remain fixed over very long periods, modeling a dynamic response between price and wage developments was not necessary. The adaptation of the BOF3 model from Finland to fit Czechoslovakian conditions provides a starting point for modeling price and wage formation for Czechoslovakia. Additionally, the growing importance of the tradables sector can be modeled within this framework.

Harald Hruschka presents his experience with missing data from market response modeling. Another interesting feature of the data in this area is its low variability due to a small variation in relevant marketing instruments. Because subjective estimates are very likely influenced by motivational and cognitive biases, a strictly empirical drawing of information from several sources is preferable. Marketing response models concentrate on using subjective indices, decision calculus, and econometric techniques either of the classical or the Bayesian sort, to combine extraneous information from already published studies with the available data for a specific problem.

II. The Time Series Approach: An Overview

Another problem that often occurs with data is the low measurement frequency. When high frequency data are needed for a particular analysis, this is a serious drawback. To bypass this problem John Chipman and Beverly Lapham compare interpolated series with realized and simulated series. In their paper they provide an empirical evaluation of interpolation methods with regard to discontinuities between observations at low frequencies and negative interpolated values. They show that non-negativity constraints and Bayesian methods can improve the results of standard interpolation methods.

Albert Jaeger and Robert Kunst point out problems that may arise for users of data that has already been interpolated. For example, in the construction of pre-war national accounts data for the USA, linear interpolation methods were a common practice. On the other hand, post-war data are based on a more extensive measurement process. Different test results for persistency measures between pre- and post-war data can be shown to be created by the interpolation procedure. The crucial variable that determines the quality of linearly interpolated data is the variance ratio between the unobserved and the underlying series. Incorrect estimates of this ratio will disturb the time series characteristics of interpolated data.

The estimation of time series models for high frequency data of the Polish economy forms the topic of the contribution by Nina Łapińska-Sobczak. Monthly and quarterly data for macroeconomic variables of the Polish economy are now available from 1988 onwards. She suggests using this information for simple time series extrapolation methods because a nonhomogeneous economic policy environment makes structural modeling difficult. Adaptive models like the Brown exponential smoothing model account for irregularities at the end of the sample period and put a higher weight on newly available information. Moreover, seasonal data can be modeled within this framework. An application to eight macroeconomic and financial indicators gives promising results.

For forecasting short and seasonally unadjusted data Baldev Raj and Thomas Url propose a structural time series modeling approach. Within an uncertain environment of unstable economic relations univariate methods are preferable over multivariate time series and econometric methods on an a priori basis. The basic structural time series model imposes structural components associated with the trend, the seasonality, or the irregular component of a time series. The question raised in this paper is whether adding structure to a univariate time series model helps overcome problems with data availability in an empirical sense. A comparison with unrestricted time series models estimated for short samples from Canada and Austria mimic the scenario in Eastern Europe and provide a guideline for the quality of this approach.

III. Case Studies: An Overview

The Polish foreign currency market between 1988 and 1989 is analyzed by Wojciech W. Charemza and George Norman. This was the most highly developed foreign currency market in Eastern Europe before the legalization in 1989. The authors develop a theoretical oligopoly model and its empirical implications on the price setting behavior in several local markets. Mobile sellers may enter or leave a particular local market but high transaction costs for purchasers prevent them from traveling across local markets. The advantage of the derived econometric model is that only price data for foreign exchange rates are needed. Given the enormous problems due to lack of data, this is considerable progress. According to the econometric tests, local foreign exchange markets can be treated separately.

In his paper on the energy sector in the Soviet Union Franz Wirl uses a partial equilibrium approach to make a forecast for energy consumption and supply since the dramatic political break. A specific problem in modeling the energy sector for former centrally planned economies is the completely distorted price structure. Absurd prices for energy automatically lead to inefficient use of energy and now there is a potential for energy conservation from several sources. These are the contracting of basic industries, the improvement of highly outdated production technology, a better motivation of the staff, and the assistance from international oil companies. Projections up to 1995 are made by a logistic fit and other extrapolation methods.

Concluding Remarks

The papers collected in this volume provide some solutions to problems associated with the statistical analysis of data from transition economies. The lack of accuracy and the shortness of data samples from such countries violate the usual preconditions of statistical or econometric analysis. Despite these difficulties, we conclude that any serious forecasting exercise must rely on objective methods to facilitate communication and critical proofs. We hope that this volume will be of help for researchers in this area and welcome any critical comments.

Part I

The Econometric Approach

Chapter 1

Problems of Estimation and Forecasting of Financial and Monetary Indicators in the USSR

Andrei I. Kazmin

1 The Main Factors of Distortion of Financial Data in Non-Market Economies

The drastic shortage of reliable data, needed both for realistic analysis of the current economic situation and for elaboration of an effective anti-crisis policy, is one of those numerous problems of transition which affects more or less equally all the post-socialist countries including the Soviet Union. Knowing the functional framework of the Soviet-type "planned" economies, one has to assume that the present lack of credible economic information is not at all surprising.

It would be enough just to mention that the official statistics in the countries of the former Eastern bloc for long years had been used mainly to present the achievements of the planned economy rather than to show the objective results of its functioning. That is the basic reason why the available official statistics failed to meet the unexpected demand for truthful information on the current economic situation and its perspectives.

Without any exaggeration one can speak about the existence of the general system of distortion of economic data on micro- and macro-level, which has consistently developed under the influence of false ideological and political priorities. The impact of purely pragmatic group interests was even more stronger – for example, the volume of the bonus fund of state-owned enterprises depended on the level of fulfillment of the plan directives, no matter how large would be the costs under normal market conditions. However, the direct falsification of primary economic data is not the only source of unreliable statistics in planned economies.

A much more important component of constant distortion of economic information is the basically wrong methodology of processing of statistical data, especially when it is presented in money equivalent. Since the early 1930s up to nowadays the estimation of the Soviet GNP or of the volume of production in separate branches of the economy is fulfilled in current roubles without correction to wholesale price index. It is worth to mention that this index was not actually calculated since price stability was traditionally considered to be the basic advantage of socialism. Only the elimination of the price factor, according to estimations of the Soviet economists G. Khanin and V. Selunin, resulted in 15-fold enlargement of the real volume of the national income of the USSR in 1985 compared to its level in 1928.[1]

This example quite fits the standards of the Orwell's Ministry of Truth though it would be hardly true to evaluate this evident desinformation just from common sense.

The degree of unreliability of macroeconomic indicators would not appear so shocking if it is properly taken into account that in this case estimations fulfilled in monetary terms are used for the evaluation of the system which is based on non-monetary planning, distribution and exchange and where money itself does not perform the key monetary functions. Due to this major reason it is absolutely impossible to achieve even minimal reliability of monetary estimations related to the non-market system.

In this way it ought to be underlined that the most unreliable data in planned economy refers to its monetary and financial system. The arguments supporting this statement are the following. First of all, unlike other kinds of economic information financial and monetary statistics is severely affected on two levels – the general data misrepresentation caused by the non-market character of the economy is enforced in this case by the distorting influence of the deeply deformed monetary and financial system which is totally adjusted to the needs of the planning-and-distribution mechanism.

Secondly, the low quality of financial statistics in planned economies is the direct result of the traditional secrecy in this sphere. Even unreliable financial

[1]Novy Mir, 1987, No.2, p. 192.

data were not able to make a proper image of the planned economy, some indicators still could tell the truth. It is not a casual coincidence that in the USSR detailed financial data became unavailable simultaneously with the beginning of the first five-year plan in 1928. Since that time nearly all indicators characterizing in absolute figures the current state of money circulation, banking, public finance and balance of payments had become a top secret information. Until the last years the available statistics contained only general figures of the state budget revenues and expenditures, volumes of indebtedness of enterprises and collective farms to the state banks and amounts of household's deposits with the saving bank.

The extreme scarceness of official statistics made it impossible to verify the available data and thus it helped to prevent the unwanted outflow of economic information.

One American sovietologist was quite right with the statement that "Russians not-willing to show us the temperature of their economy have broken a thermometer in their own house".

The traditional double standard in data processing-one for the public and the other for the government structures had a tremendous negative impact on the general level of economic research and contributed to the unreliability of statistical information. One can not exclude also the influence of the professional gap which remains till nowadays since the years of big terror in 1930s. After Stalin had qualified the studies of well-known economists as a "senseless game with figures" nearly all the best scientists and scholars in this field became the victims of horrible repressions. The formerly famous research centers of econometrics among them: The Conjuncture Institute, headed by Prof. N. D. Kondratjev and The Conjuncture Department of the State Planning Committee, headed by V.G. Groman, had ceased to exist. Alongside the Central Statistical Agency was transformed into a subordinate body of Gosplan. After all those strokes the former highest culture of econometric studies had been lost and the possibility of it's revival still looks very doubtful. The inherited pathological secrecy is not yet abandoned either.

Glasnost reached the financial statistics only in the last few years. In 1988 came the official acknowledgement of the fact of the budget deficit followed by publication of its amount. One year later the amount of the state debt became also available to the public. In 1990 the figures of the emission of money in cash were revealed and in 1991 the total volume of cash money in circulation was also made public.

However, the list of secret information still contains the balance of the State Bank, the balance of payments of the USSR, the amounts of gold and hard-currency reserves, the volume of external indebtedness, the structure of money in circulation, etc.

Consistent opening of the financial data is extremely necessary for elaboration of a realistic program of transition to market economy. And the earlier this process would succeed the better, but it will be a great mistake to think that from this time on actual problems, caused by the unreliability of economic data needed for decision making, would be solved. As it was already explained earlier, the remaining secrecy of financial statistics can never be considered as a guarantee of its credibility. Even absolutely honest information is inevitably distorted under the conditions of the non-market economy and its deformed monetary and financial system. The analysis of the estimations of the amount of the state budget deficit of the USSR give a good example of such kind of distortion.

2 Unreliability of the Official Budget and Evaluation of Its Real Deficit

Complicated calculations are not necessary to become suspicious on the unreliability of the official reports about the volume of the budget deficit. One has to pay just more attention to the fact that since 1988 alongside with the yearly statements of the figures of the budget deficit the final reports adopted by the Supreme Soviet which are reproduced in statistical year-books contain a budget with a surplus of revenues over expenditures.

Thus, according to the official statement of Goskomstat the amount of the USSR budget deficit in 1988 (the first year when the deficit was revealed) equaled 80.6 bn roubles[2]. At the same time the Supreme Soviet of the USSR adopted the report on the state budget in 1988 with a 9.5 bn roubles surplus[3]. This paradox was reproduced in 1989, when the 80.7 bn roubles budget deficit was accompanied again by a 11.1 bn roubles surplus[4].

To find some explanations to this phenomenon, one has first to answer what is the source of surplus revenues while the budget is fulfilled with deficit. Today the Ministry of Finance does not make a secret from this source any more – the budget deficit is covered by so called accumulated funds, mostly

[2]See, Narodnoe Khozyaistvo v SSSR/Statistical Year-Book, 1989, Moscow, p. 614.

[3]See, Ekonomicheskaya Gazeta, 1988, November, No.47, p. 21.

[4]See, Narkhoz, 1989, p. 613–614.

represented by the direct loans provided by the State Bank[5]. The volume of these funds used for financing the budget deficit is presented in the following table:

Table 1.1:
The Volume of Accumulated Funds and of the State Budget Deficit of the USSR (bn roubles)

	1985*	1986*	1987*	1988	1989
Accumulated Funds	18,0	47.9	57.1	90.1	91.8
Budget Deficit (official report)	13.9	45.5	52.5	80.6	80.7
Surplus Financing of the Budget Deficit	4.1	2.4	4.6	9.5	11.1

* announced in 1989[6].

Knowing the volumes of this surplus one still needs an explanation what for the budget deficit is financed in excessive amounts. The reasons for such tricks before the perestroyka were quite clear – the surplus budget served to keep the image of the stability of the Soviet financial system. The situation became absurd since 1988 when this myth was left and the newly acknowledged budget deficit began to coexist with its surplus.

The analysis of the forms of inter-connection of the budget and the State bank's loans gives a rather pragmatical solution of this puzzle. The case is, that in the framework of the plan-type monetary system the State Bank is obliged to provide credits to the Ministry of Finance during the whole financial year in order to cover the constant shortage of collected revenues. These loans are provided through direct money emission without any sort of collateral and fixed terms of repayment.

The precise demand of the budget for such loans can be finally figured only after the end of the financial year when the exact amount of actually collected revenues would be known. As a result the budget obtains more bank loans during the current financial year than is really needed and the yearly remainder constantly shows the surplus of budget incomes (including accumulated funds) over the volume of expenditures. But the financial alchemy does not stop at this point. According to the regulations which remain valid until now the budget surplus produced by excessive bank loans is legally transformed into budget deposits with the State Bank. The latter uses it for outstanding new

[5]Beside the loans provided by the State Bank, the accumulated funds include the yearly surplus of the houshold's deposits with the Saving Bank and the reserves of the State Insurance Company.

[6]See, Narkhoz, 1989, p. 614.

credits. That mechanism provides automatic refinancing of the State Bank's loans into budget incomes and vice versa. So, the more credits are provided to the budget in the current year by the State Bank, the more loans it would be able to provide in the next year.

The impact of this single element of the abnormal monetary and financial system results in artificial enlargement of budget incomes and lowering of the real volume of the deficit. Alongside it helps to exaggerate the amount of deposits in the balance of the State Bank and accordingly to multiply the volumes of credit and monetary emission.

Thus in order to obtain more reliable data on the budget deficit it is necessary to exclude from the volume of incomes the whole amount of the "accumulated funds". Secondly, the budget incomes must be also reduced by the amount of the treasury bonds sold to households and the so called "other sources of incomes". Actually the latter do not exist since this figure represents the duplicate counting of the revenues temporarily withdrawn from the state budgets by the federal budget and returned back within a year[7]. The results of these calculations could be shown in table 1.2.

While the official budget report of 1990 is not yet accepted, the volume of the budget deficit could be reconstructed according to single data available from various publications.

Thus, even the primary correction of the available data shows that the officially reported volumes of budget deficit are actually dumped by 30 % at minimum and by 3-fold at maximum. Surely the real figure of data distortions would be even greater if one takes into account the indirect channels of enlargement of budget incomes through guaranteed payment of turnover tax or agricultural subsidies. The inclusion of the export earnings into budget incomes contributes to minimization of the reported budget deficit.

[7]See, J. Liberman, Force-major of inflation, Druzhba Narodov, 1990, No. 8, p. 230.

Table 1.2:
The Evaluation of the Corrected Volume of the Budget Deficit (bn roubles)

		1985	1986	1987	1988	1989
1.	Total Budget Incomes (Official Report)	390.6	419.5	435.5	468.5	493.7
2.	Excluded:					
	– Accumulated Funds	18.0	47.9	57.1	90.1	91.8
	– Public Treasury Bonds	1.4	1.9	1.9	2.0	2.1
	– "Other Sources" (Duplicate Counting)[8]	25.6	25.3	23.8	26.6	29.8
	– Totally:	45.0	75.1	82.8	118.7	123.7
3.	Corrected Budget – Incomes (1–2)	345.6	344.4	352.7	349.8	370.0
4.	Total Budget – Expenditures	386.5	417.1	430.9	459.5	482.6
5.	Corrected Budget – Deficit (3–4)	40.9	72.7	78.2	109.7	112.6
	– in percent to GNP[9]	5.2	9.1	9.5	12.5	12.3
6.	Official Budget Deficit	13.9	45.5	52.5	80.6	80.7

Table 1.3:
The Evaluation of the Budget Deficit in 1990

1.	Reported Budget Incomes (without actually sold government obligations of 1990 – 58.9 bn roubles.[10])	452.0[11]
2.	Excluded:	
	Annual Surplus of Households Deposits (accumulated funds)	42.9[12]
	– Treasury Bonds Sold to Households	6.5[13]
	– Other Incomes (Duplicate Counting)	30.6[14]
	– Totally:	80.0
3.	Corrected Budget Incomes (1–2)	372.0
4.	Total Budget Expenditures	510.1[15]
5.	Corrected Budget Deficit (3–4)	138.1
	– in percent to GNP	15.2
6.	Official Budget Deficit	58.1[16]

[8]Estimated according to the data presented in the Statistical Year-Book: Narkhoz, 1989, p. 611.

[9]Estimated, see: Narkhoz, 1989, p. 6.

It is extremely difficult to calculate the influence of above mentioned factors not only because of the visible unreliability of the major part of the available data but mostly because of the growing uncertainty about the future constitutional basis of the Soviet Union and its budget system in particular. Even the short-term forecasts of the financial situation are very vague due to the rapid decline of production and the expected separation of 6 republics from the newly formed Union.

Nevertheless several attempts were undertaken to evaluate the impact of the drastic price increase, which took place in the last April, on the budget deficit. The Soviet economist S. Alexashenko argued that its total volume would reach 134.7 bn roubles[17]. by the end of the year what surely seems to be closer to the truth than the ridiculous officially planned amount of 26.6 bn roubles[18]. According to the statement of the State Bank it took only three months to overcome this yearly limit by 4.5 bn roubles[19]. Under these circumstances the union government lost interest to minimization of the budget deficit and declared that its volume may rise to 200 bn roubles by the end of 1991[20].

In spite of all the problems of evaluation and forecasting of the real budget deficit this task can hardly be regarded as the most complicated one, while the budget statistics was always the only available source of financial information. The situation was quite opposite with the data related to monetary circulation, regarded as the main object of the most strict censorship.

---

[10]See, Izvestia, 27. May 1991, p. 2.

[11]See, Ekonomika i Zhizn, 1991, January, No. 5, p. 9.

[12]Ibid., p. 10.

[13]See, Pravda, 5. December, 1990,p. 2.

[14]Estimated according to data published in Ekonomika i Zhizn, 1991, January, No. 5.

[15]Ibid, p. 9.

[16]Ibid.

[17]See, Nezavisimaya Gazeta, 6. April 1991, p. 2.

[18]See, The law on the Union Budget of the USSR for 1991, art.5; Izvestia, 15. January 1991.

[19]See, Izvestia, 3. April 1991, p. 1.

[20]See, The Report of the Prime-Minister V. Pavlov on the Session of the Supreme Soviet of thew USSR; Izvestia, 23. April 1991, p. 2.

3 Primary Estimations of the Soviet Money Aggregates

Due to the long experience with totally unavailable monetary statistics there are no reliable figures of the volume of money in circulation and its single aggregates. Without this kind of data it is hardly possible to elaborate a really effective anti-crisis program for the Soviet economy.

Thanks to the progress of Glasnost the problem of making the money data available is likely to be solved very soon, since V. Gerashenko, the Chairman of the Board of the State Bank of the USSR has promised to lift the existing bans on banking information until the end of the year[21].

But it is much more harder, like in the case with the budget, to eliminate the distorting influence on the money indicators produced by specific elements of the non-market monetary system itself.

First of all, beginning the analysis of money indicators in non-market economy one must fully take into consideration such a basic principle of the Soviet monetary system as the strict separation of circulation of cash money and money on bank accounts. This barrier between cash and accounting money had been erected in 1930 when all legal economic agents were subordinated to the planning-and-distribution mechanism.

In practice this artificial split (by market terms) of the money turnover is implemented through absolutely different methods of regulation of cash and accounting money. It would be correct in this case to speak about the co-existence of two autonomous monetary systems: one for cash and the other for accounting money.

This explanation helps to understand why the Soviet official dictionary and statistics qualify money in circulation exclusively as money in cash, as well as money issue(emission) relates only to issue of cash. This approach is quite logical in the framework of the planning-and-distribution system since only by using cash money its owner has at least some freedom from administrative control over his spendings. For this reason traditional monetary policy in a non-market economy prescribed severe restrictions of the volume of cash in circulation. Besides, cash circulation attracted much more administrative attention because it embraced only the household's incomes and payments and in this way it influenced directly the state of consumer market and accordingly the social temperature in the whole country. Under these conditions the Soviet economic studies in that field were usually narrowed to the evaluation of the unsatisfied payment demand of household.

---

[21]See, Kommunist, 1989, No. 16, p. 26.

The accuracy of such evaluations is rather doubtful, since their authors presented absolute figures of the volume of so called "hot money" without revealing the implemented methodic of calculation. While the majority of necessary data remained unavailable there was no other way but to look for some indirect indicators related to money circulation. Therefore it is no wonder why the specific competition of speculations over real amounts of money overhang have been always so popular among the Soviet economists.

The range of evaluations has become tremendous especially during the last two years. They differed from the minimum of 70 bn roubles[22] estimated by N. Petrakov to the officially presented medium of 150 bn roubles[23] up to the maximum of 954 bn roubles announced by J. Liberman[24].

Putting aside its extremeness, the last figure is the most interesting one because unlike all other estimations it is based on an available method of calculations. The essence of this method is rather simple but convincing. Comparing direct and indirect data on household yearly incomes and spendings, J. Liberman came to the conclusion, that even current payment demand constantly exceeds the volume of supply of consumer goods and services. In this case unsatisfied demand accumulated during the current year has no chances to dissolve afterwards. Therefore the yearly surpluses of money overhang are constantly enlarging the amount of accumulated unsatisfied demand which nowadays covers nearly the whole volume of household savings both in cash or in deposits with the Saving Bank.

To estimate the total amount of accumulated money overhang (M_n) no complicated econometric instruments were needed. It was enough just to integrate the yearly balances of household's earnings (E) and payments (P):

$$M_n = (E_0 - P_0) + (E_1 - P_1) + ... + (E_n - P_n); \qquad (0, 1, .., n - \text{appropriate year})$$

After the volume of household's deposits with the Savings Bank (Sd), which is available and reliable, is subtracted from the total amount of money overhang, the remainder shows the sum of household cash savings (Sc):

$$Sc = M_n - Sd.$$

[22]See, People' Deputies of the USSR: Economy Today and Tomorrow, Moscow, Moskovsky rabochij, 1990, p. 77.

[23]See, Ryzhkov, N.I., The Government Report to the 2nd Congress of People's Deputies of the USSR, Moscow Politizdat, 1990, p. 27.

[24]Liberman J., The Force-Major of Inflation, Druzhba Narodov, 1990, No.8, p. 235.

In its turn the latter is the basic data to evaluate the volume of cash in circulation and the yearly amounts of cash issue. The results of defined calculations, accomplished by J. Liberman can be arranged in table 1.4:

Table 1.4:
Estimation of Accumulated Money Overhang (bn roubles)
(end of the year)

		1985	1986	1987	1988
1.	Annual Surplus of Unsatisfied Payment Demand ($E_n - P_n$)	93	100	107	123
2.	Accumulated Unsatisfied Demand (Money Overhang) (M_n)	260	360	467	590
3.	Household Saving Deposits (Sd)	221	243	267	297
4.	Household Cash Savings (Sc)(2-3)	39	117	200	293
5.	Current Payment Demand (Household Annual Earnings-Taxes excluded)	418	n/a	449	487
6.	Total Payment Demand (2/n-1/+5)	–	–	809	954

In spite of the evident logic of described estimations there are several serious methodological mistakes that caused the tremendous exaggeration of final results. First of all, it is not correct just to add directly the annual volume of current payment demand to its accumulated remainder (turnover data in this case is mixed with balance data). Moreover the received figure certainly cannot be equal to the volume of money in circulation. Secondly, calculating the annual surplus of unsatisfied demand one must certainly take into account all the household payments: beside the amount of retail sales and taxes there are also paid services, government bonds, insurance payments etc.

After the necessary corrections are done the above displayed volumes of annual surpluses of money overhang appear to be reduced by 2–3 times. This proves the reliability of the newly published data on this matter. Using these statistics it is possible to give a more accurate evaluation of the volume of household's cash savings (in table 1.5).

Table 1.5:
Corrected Evaluation of Household's Cash Savings (bn roubles)
(end of the year)

		1986	1987	1988	1989
1.	Annual Surpluses of Unsatisfied Payment Demand[25]	28.0	32.0	41.9	61.8
2.	Annual Surplus of Saving Deposits	22.0	24.1	29.8	41.1
3.	Including Surplus From Interests Paid on Deposits	5.8	6.4	7.0	8.0
4.	New Bonds Sold to Household	1.9	1.9	2.0	2.1
5.	Annual Surplus of Cash Savings [1-(2-3)-4]	9.9	12.4	17.1	26.6

As it can be seen from the Table the accumulated cash savings surplus during the taken 4 years is evaluated to 66.0 bn roubles. Taking into account that the inflationary growth of earnings has begun in the middle of 1980s, it can be accepted that the cash savings accumulated before the mentioned period were relatively small. Adding the volume of accumulated cash savings to the amount of cash normally demanded by the economy – (35-50 bn roubles)[26] it is possible to define the total volume of cash money in circulation, which by the beginning of 1990 equaled to 120 bn roubles.

This evaluation coincides with the results received by western experts[27] and is very close also to the newly opened official data what may be regarded as a certain assurance of the reliability of the latter. According to statistical report the volume of cash savings has increased from 104.7 bn roubles by the end of 1989 to 132.7 bn roubles in the next year[28]. Knowing the dynamic of annual change of amounts of cash emission it is possible to calculate the volumes of cash savings in previous years. After adding cash holdings of legal persons the total sum of cash money in circulation (Mo) can be finally calculated. The annual volumes of this aggregate are the following:

	1987	1988	1989	1990
Household Cash Savings (end of the year)[29]	78.7	86.7	104.7	132.7
Cash in Circulation(Mo) (estimated)	83.0	92.0	111.0	140.0

[25]See, Narkhoz, 1989, p. 76.

[26]See, Pavlov V.S. Finance is Our Main Problem, Moscow, Financy i Statistika, 1990, p. 274.

[27]See, Ofer Gur, Macroeconomic Issues of Soviet Reforms, Paper presented at the 5th Annual Conference on Macroeconomics, Cambridge MA, April 1990, p. 40.

[28]See, Ekonomika i Zhizn, 1991, January, No. 5, p. 9

When the estimation of this formerly most secret data is completed it becomes possible to calculate other money indicators and to arrange them due to the actual level of their liquidity within the specific framework of the Soviet monetary system. Definite peculiarities of this system had been ignored by the elaborators of the "500-Days"-Program, who were the first to give the draft picture of the structurized Soviet money volume[30].

Table 1.6:
Evaluation of the Soviet Money Aggregates (bn roubles)
(by the end of the year)

		1987	1988	1989	1990
I.	Household Liquid Assets				
	1. Household On-Call Saving Deposits	159.2	176.1	201.8	227.3
	2. M1: M0 + 1	242.2	268.1	312.8	367.3
II.	Household Term Assets				
	3. Household Term Saving Deposits	107.7	120.6	135.0	152.0
	4. Deposit Certificates	–	–	1.0	1.4
	5. State Bonds	1.9	2.0	2.1	4.0
	6. Insurance Policies	10.7	11.2	11.7	12.0
	7. Total (3+4+5+6)	120.3	133.8	149.8	169.4
	8. M2: M1 + 7	362.5	401.9	462.6	536.7
	Ratio: M2 to Commodity Stock in Retail Trade	5.534	6.221	7.183	8.332
III.	Financial Assets of the Public Sector				
	9. Current Accounts of Legal Persons	92.3	110.8	117.0	134.0
	10. Budget and State Insurance Reserves (evaluated)	15.0	16.0	18.0	20.0
	11. Social and Investment Funds	67.4	102.5	108.9	125.0
	12. Amortization Funds	13.3	12.7	15.0	21.0
	13. Total (9+10+11+12)	188.0	242.0	259.0	300.0
	14. Net Indebtedness to Banks	429.0	398.8	383.7	354.4
	15. M3: 13+14	617.0	640.8	642.7	654.4
	Ratio: M3 to Stocks of Capital Goods	1.316	1.340	1.385	1.410

Sources: Narkhoz, 1989, p. 94, 611, 617;
Finance of the USSR, Statistical Yearbook, 1989, p. 39, 47, 52;
Dengi i Kredit, 1990, No. 10, p. 72–74, 1991, No. 3, p. 6; 1987, No. 11, p. 55.

[29]Estimated according to: Pravda, 28. January 1990, p. 1; Kommersant, 1990, No. 42, p. 22.

[30]See, Transition To Market. Concept and Program (in Russian), Moscow, August 1990, p. 227.

Due to the remaining barrier between cash and accounting money it would be not correct just to combine household and public money assets into one aggregate according to accepted western standards. Besides such separation helps to see that the money overhang on the consumer market at present is several times bigger than on the newly formed market of means of production. This situation must be fully taken into account in the now elaborated program of the radical reform of the entire monetary and financial system. Though actually this paradox is explained by the sustained administrative control over prices, earnings and payments of state-owned enterprises. It would surely vanish along with the coming marketization of the public sector. Only then the problem of credibility of financial data would be solved and evaluations of money indicators would become quite reliable for the sound economic policy.

Chapter 2

Macroeconomic Forecasting in the Transition Period – The Case of Hungary*

JUDIT NEMÉNYI

1 Introduction

Since the beginning of the seventies several macroeconometric models with different assumptions have been elaborated for policy evaluation and forecasts in Hungary. Following the first attempts showing the usefulness of econometric approach in CPE (Halabuk et al. [1973, 1976], Simon [1979]), the econometric modelbuilding and forecasting was closely related to the national planning activity (Hunyadi et al. [1979], Hulyák [1982, 1986, 1989], Hunyadi et al. [1985]), although the econometric models have never become an integrated part of the standard planning methodology based mainly on input-output analysis and on consistency testing of estimations made by planners.

The main reasons why econometric models were rejected or treated with suspicion had been changing a lot. In the seventies, firstly, the basic concept of the econometric approach was rejected by saying that economic agents' behavior in the CPEs was determined by planning targets. Then, with loosening central direct control, the lack of theoretical background, the shortness of

*This paper was prepared within a research program dealing with 'Problems of Forecasts in the Transition Period' and financed by OTKA (National Scientific Research Fund (No. 293)).

time series available and the missing data problem made difficult to produce reliable results. In the eighties, in the circumstances of the emerging market processes and of the deepening crisis, the validity of macroeconometrics for CPEs was disputed either because of some political reasons or because of their underlying theory. The forecasts made by econometric models containing estimated behavioral equations for the different economic agents had often contradicted to the planned indicators or highlighted some inconsistences of planning, which could not be taken into consideration because of the "anchors" determined by the ruling party (e.g. consumption price index, budget deficit etc.). The grounds of stochastic behavioral equations had not been queried anymore but the underlying theory of the specification was rejected very often. The specification of macroeconometric models have been based generally on the adaptations of theories elaborated for market economies, the validity of which in a CPE had been criticized. The theory of shortage (Kornai [1980]) reflecting the CPEs' characteristics did not accept any symmetry between market economic phenomena and CPE's mechanisms (such like analogy between repressed inflation in a fix-price capitalist economy and the "suction system" of socialist economy, validity of disequilibrium approach on consumers' good markets etc.), but it could not provide a general macroeconomic framework suitable for macroeconometric verification.

Considering empirical results we can agree with Portes's assertion concerning the validity of macroeconometrics for CPEs: "There are no general a priori guidelines for the essential elements of a CPE macro model. All simplifying assumptions are 'wrong', so one chooses a minimal coherent set sufficient for one's objectives. ... One could query the basic strategy of applying to CPEs the neoclassical macro theory, in its quantity-rationing version, as well as disequilibrium econometrics. I believe, however, that they and a simple monetarism are often better suited to CPEs than to comparable market economies" (Portes [1984] p. 18). In Hungary the econometric modeling in CPE was struggling against several theoretical and methodological difficulties as can be seen from the above. Since 1990, the beginning of the transition period from the CPE to a market economy, a number of new problems have to be considered.

In this paper, first, the main characteristics of the Hungarian CPE's macroeconometric models will be presented briefly. The second part deals with the current and expected economic changes and transformation of the information system allowing of their monitoring, both decisive from the point of view of the econometric model building. Finally an attempt will be made to outline the possibilities of forecasting in the transition period.

2 Macroeconometric Models of CPE in Hungary

Hungarian macroeconometric models disposable at the end of the eighties – sectoral (Hunyadi et al. [1985]) or aggregated ones (Halpern [1989], Hulyák [1989], Neményi [1990]) – focus on the real side of the economy. Domestic prices are either given exogenously (Hulyák, Neményi) or simulating the rules of the price system implemented in 1980 (which has slightly liberalized the previous central price control), they are computed using input-output tables, distinguishing centrally fixed and cost based prices. Export and import prices are derived from the exogenous foreign market prices considering export subsidies and import tariffs. Fiscal part of the models contains generally the main identities of income generation and redistribution with fix parameters.

The annual national accounting system forms the conceptual framework and the empirical basis of the models. The blocs describing real processes of the models deal first of all with the final demand and aim at the verification of excess demand on the main markets of goods (consumption and investment goods). For this purpose the different models use different approaches. In the disequilibrium model (Hulyák [1989]), taking into account the repressed inflation condition, the savings adjustment equation has been used for the consumption good market and a plan adjustment mechanism has been verified in the market of investment goods, supposing that planning activity intends to decrease the tension caused by intensive permanent excess demand. In the ERI macromodel (Neményi [1990]) business survey data have been used as proxies for the description of the excess demand. The empirical results show unanimously that due to the political efforts to increase - later on to protect – living standard of the population, in the consumption goods market an 'almost-equilibrium state' has been realized. On the contrary, in the investment goods market significant permanent excess demand of fluctuating intensity could be verified.

In the aggregated models the supply of domestic output is determined by the main production factors (labor and capital) considering technical change and capacity utilization. In the sectoral models elaborated directly for the national planning, according to the planning methodology, the input-output technological coefficient matrix has been used for the determination of the total output and imports.

The core of the models has been organized around the balance of GDP. According to the increasing international indebtedness of Hungary during the eighties, the problems of the foreign disequilibrium have been getting more and more importance in the modeling. Price, income and subsidization effects in the exports and imports have been investigated (Hulyák [1986], Halpern

[1989]). In almost each of the models the foreign imbalances (trade balance, balance of payment surplus/deficit) have generally used as tensions indicators and have been considered as proxies of central control, having a lagged feedback to the final demand. On the contrary, the state budget deficit despite of its great importance in evaluation of the country's performance, did not much figure in the feedback mechanisms of the models, which is likely due to the lack of reliable, homogeneous time series. For the same reason the fiscal relationships are described only very roughly.

Despite of improving performance of the macroeconometric models, which manifested in interesting policy evaluation results and forecasts 'acceptable' for planners, the macroforecasts based on econometric approach using time series have come to a serious than ever crisis by the beginning of the nineties. Regime changes cause always difficulties in forecasting, but the problems the transition countries are facing actually are more complex than any of the well-known techniques may provide a general solution for them.

3 Characteristics of the Transition Period and the Modeling Consequences

In order to characterize the observation period of the macroeconometric models the annual average growth rates of the main macroprocesses for the two main subperiods are given in table 2.1. Also the corresponding figures of the government program are included, covering the period of forecasts (details of the program are given in table 2.A1 in the appendix). In order to make clearer the slow down of Hungarian economic development by the end of the eighties, the time path of GDP and the domestic absorption can be seen on figure 2.1.

Table 2.1:
Growth Rates (percent)

	1965-80*	1980-88*	1989	1990	1991**	1992**	1993**
GDP	5.6	1.6	-0.2	-4.5	-3.5	3.0	5.0
Industry	6.4	1.0	-3.4	-6.0	-7.5	2.0	4.0
Agriculture	2.7	2.4	-1.8				
Households' consumption	5.7	1.3	0.4	-5.0	-5.0	1.0	2.0
Investments	7.0	-1.0	4.3	-7.0	-4.0	0.0	5.0

Source: Main economic indicators, CSO Hungary 1989,

* Average growth rates, World Development Report 1990, World Bank,

** Stabilization and convertibility (Economic program for 1991–1994), Ministry of Finance, February 1991.

If the government targets are well-founded and the shift to a moderate growth path by 1992 is not beyond the capacity of the Hungarian economy, it is evident that the constancy assumption concerning the estimated parameters - reflecting factor productivity and adjustment responses of the observation period - cannot hold. On the other hand, however, the past experiences show that during the period of the introduction of market elements into the CPE (the 1980s) the structural and efficiency changes were much slower than had been expected. For this reason even in the present situation it would be informative to compute *ceteris paribus* forecasts.

Figure 2.1:
GDP and Domestic Absorption
(1970 = 100, at 1981 constant prices)

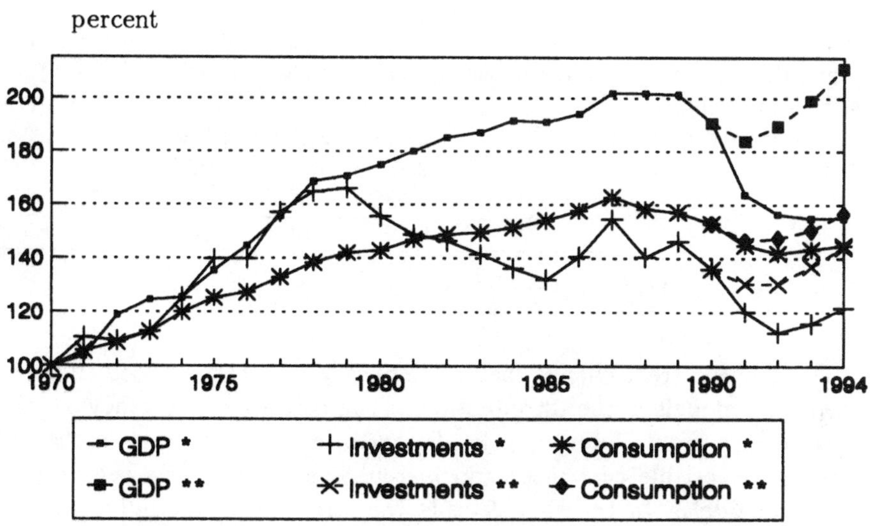

* Actuals for 1970–93, forecasts for 1994
** Government's Program published in 1990

Structural changes – model changes

There are two basic changes why actual models cannot be used for inference without any revision. Both have been influencing the operation and behavior of the firms, especially in the industrial sector, the performance of which is crucial from the point of view of rapidity and success of the transition. The first one is the change of ownership structure by privatization and creation of new enterprises. This influences the formation of every macroeconomic processes, and will probably induce behavioral, technological and efficiency changes as well. This is the case when not only the parameters but the specification of the model as well is expected to change. Figures concerning the development of new firm's establishment are given in table 2.2.

Table 2.2:
Number of Firms Classified by Institutional Form

	1987	1988	1989	1990	1993
		31 December			
Enterprises, trusts	1,352	1,377	2,399	2,367	1,130
Joint companies	302	309	327	237	145
Associations	69	78	105	201	237
Joint stock companies		116	307	646	2,375
Limited liability companies	137	451	4,485	18,317	72,897
Agricultural fishing and specialized agricultural cooperatives	1,337	1,333	1,333	1,400	8,175
Other cooperatives	2,658	2,439	2,510	2,577	n.a.
Small cooperatives	2,154	3,108	3,233	3,155	n.a.
Other	588	600	470	576	493
Total	9,597	10,811	15,169	29,470	85,452

Source: Monthly Statistical Bulletins, CSO Budapest

The other current change making impossible forecasting with the macroeconometric models is the introduction of the convertible currency settlement in the former rouble foreign trade. Hitherto, distinction between the foreign trade in convertible currencies and in rouble settlement has been crucial in all macromodels. In the rouble trade the striving for an equilibrated trade balance served as a basis of the specification up to the end of the observation period. Significant demand side effects could be verified in the trade settled in hard currencies (Hulyák [1986]), and the resulting (dis)equilibrium position in dollar has influenced the dynamic time path of the domestic absorption. Experiences of the rouble trade cut in 1990 proved that there were several 'reserves' in the enterprises to realize market reorientation, but this could be only partly forecasted on the basis of annual time series (only some demand side effects). Neither production loss effects of the rouble foreign trade decline could be captured by models including supply determined production functions.

These two fundamental changes – structural changes of the ownership system and the foreign trade – by themselves would be enough to argue that macroeconometric forecasts need some additive elements allowing to track somehow the different tendencies of future development, most of them rather gradual and not so radical like the rouble trade fall.

The share of the new and privatized enterprises, the private sector, is increasing (it is supposed to grow above 50 percent in a few years) as the weight of the former state-owned sector is declining. This 'mixed' model consists of the combination of extrapolated past and estimated/expected changes supposing varying weights. The emerging market processes – that should be decisive to promote recovery of the country – have no historical background. There are a lot of problems in data supply concerning the new enterprises. Even if the data collection problems are solved it will take some years until the new behavioral equations may be specified on the basis of time series.

Moreover, the growing competition influencing both supply and demand side and the changes of the government economic policy may likely result that neither of the parameters characterizing the state-owned sector's past operation can be taken unchanged. The question may be raised, whether the equations estimated for the past can be accepted for characterizing the operation of state-owned sector, and how long. What is the direction of parameter and/or model changes and what is the suitable information base for their identification?

Long-term – short-term

The rapidity of the adjustment is an other crucial point of the development in the transition period. Because of the availability of statistical information the macroeconometric models have been estimated on the basis of annual data, however, for tracking dynamic adjustment quarterly data would be preferable. As market forces are spreading, for business cycle forecasting, the elaboration of suitable indicators would be very important as well. The specification of the macromodels would need a general revision considering recent developments in econometrics (e.g. in the cointegration theory). The identification of long-term relationships of macroprocesses and the re-specification of the models allowing for the description of short-term effects would increase the accuracy of forecasts.

Missing blocs

There are some 'white fields' in the models of CPE. Unemployment can be considered as such a problem, brought about by market economy. The full employment system maintained in the CPE, motivated mainly by political aims, was reflected in the models treating labor as exogenous and not dealing with unemployment rates, although the notion of the 'hidden unemployment' was well known for long. The decline in the output and in the rouble exports coupled with the bankruptcy of several state-owned large enterprises have resulted an unemployment rate of about 3 percent in 1990, which is not very high in international comparison, but it is expected to increase as a consequence of the competition inducing factor productivity changes in the whole industry.

The liberalization process (concerning prices, wages, imports etc.) has been one of the necessary conditions of the transition, but at the same time it can be considered as one of the sources of the accelerating inflation problem (about 30 percent consumer price increase in 1990). The inflation forecasts are of primary importance from the point of view of the stabilization.

In the models of the CPE financial blocs contained accounting identities of income redistribution using parameters of the economic regulation system (direct and indirect taxation, subsidization). For the lack of capital markets there was no reason to suppose any kind of portfolio choice in the investigation of the economic agents' behavior, and the problem of financial integration has been avoided. Since 1988, in Hungary, the main institutional preconditions for a market oriented decision making of economic agents (the two tier banking system, Budapest Stock Exchange) have been established, which makes it necessary to complete forecasting models with a description of different agents' behavior (enterprises, financial institutions, households etc.) in financial markets. Also the effects of governmental monetary policy should be taken into account in order to investigate the effectiveness of increasing monetary and fiscal rigor, which are supposed to be one of the main elements of the stabilization program. The transmission mechanism towards the tight monetary policy – said to be pursued for years – was beyond the scope of the models. Hitherto this has not been regarded as a real inadequacy because the positive effects of a tight monetary policy have been partly vanished due to involuntary trade crediting of many Hungarian enterprises, which can be considered a special type of money substitutes.

How is it possible to produce macroforecasts?

Taking into consideration all the above mentioned changes in Hungary, actually there is a certain skepticism concerning the possibility of macroeconomic forecasting with econometric models. As the international experiences show (see in Wallis [1989]) the crises in forecasting activity – such like failures in the mid-seventies – could be explained by the inadequacies of the models (oversimplified treatment of the supply factors and expectations) and the revision of the macroeconometric models (concentrating on the financial sector and monetary policy, more detailed analysis of the supply side and replacing adaptive expectations by 'model-consistent' ones) led to the reappraisal of forecasting by econometric models. Given the never-known complexity of transition problems to be solved, it is very unlikely that in the near future macroforecasts of transition countries can be produced as an outcome from any kind of simultaneous econometric model consistently estimated using historical data.

In the transition period external and internal equilibrium, however, are of primary importance from the point of view of the stabilization, as well as the adjustment process and its dynamics. For this reason, we think, that the macroeconometric modeling must not be abandoned, but should be kept as a framework of forecasting. However, the details of forecasting systems should be based on theories and methods considering the characteristics of the emerging processes. It should be investigated which are the processes – if there are any – and whether validity of time series based estimation can be maintained henceforward in the transition period.

In the next part of the paper an attempt will be made to outline a possible transformation and extension of the macroecomonetric model used for regular forecasts at the Economic Research Institute (Neményi [1990]) in order to make it suitable for forecasts in the transition period.

4 Forecasting System in the Transition Period

The present version of the ERI (Ecònomic Research Institute) macromodel[1] is a small size (25 equations) econometric model estimated on the basis of annual time series for 1970–1989. Behavioral equations of the model relate to the aggregate supply (domestic and imports) and to the components of final demand (consumption, investments, exports). Prices are exogenous and disposable income is derived from GDP at current prices using exogenously given parameters. The estimated equations are given in the appendix and the main interdependencies of variables can be seen in the chart below.

[1]This paper was presented in 1991. The version of ERI model described in the paper was estimated on the basis of time series for 1970–1990. The economic development of the early nineties differed a lot from what had been and could have been expected (deeper recession, higher budget deficit etc.). Main features of the transition period have been recognized gradually and besides the internal restructuring problems external factors (general recession of developed countries, wars etc.) contributed to the delay of recovery. As a consequence forecast models have to be respecified in order to take into account the segmented trends in the economic development. This would be the subject of an other paper. Here all we can do is to compare the ERI model's forecasts made in 1991 with the actual processes (see actual data in the appendix and figures).

Figure 2.2:
The Structure of the ERI Macromodel

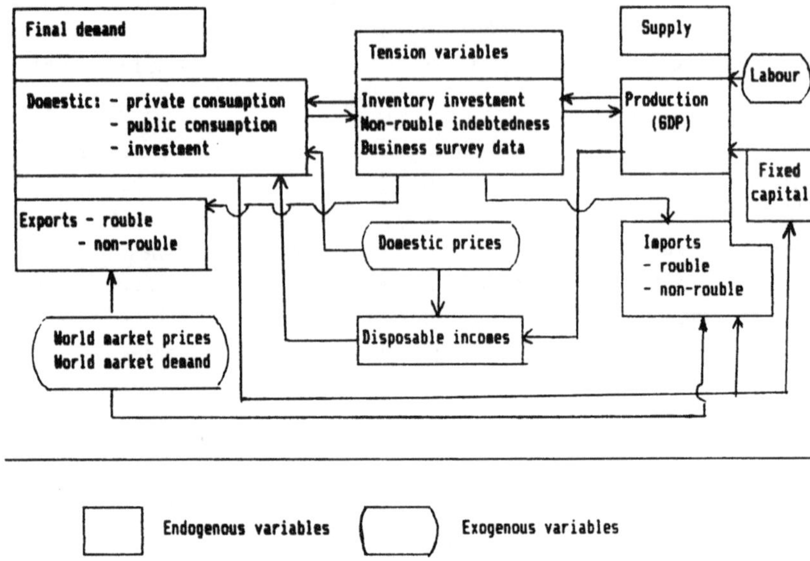

The specification of the model is based on the observation that in the eighties economic development in Hungary can be characterized by a modest average growth rate of domestic supply and absorption while foreign convertible debt has increased significantly (debt/GDP over 60 percent by 1990). In order to solve the permanent current account problems different measures have been introduced to reduce domestic demand and to promote the creation of a foreign trade surplus. In the model, according to the excess domestic demand hypothesis, the households' consumption and the firms' investments are determined by demand factors (incomes and prices) and by policy variables representing central efforts to reduce domestic demand. The inefficiency of market mechanisms and various restrictive measures have created a lot of bottlenecks in the production, while the low growth (later on stagnation) of output has further constrained the possibility of an increase in domestic absorption. The aggregate supply is described by main primary production factors (capital and labor) and by variables representing tensions in the production process.

In the model there are three types of variables reflecting tensions generated by internal and external imbalances. The change in the inventory investment is derived from the GDP balance identity assembling supply and demand effects as well. The growth of the convertible indebtedness depends directly on the trade balance in the current year and indirectly on the foreign trade performance of previous years through the financial needs for debt service obligations. Since 1968 regular business surveys have been carried out with state-owned enterprises by the ERI concerning different fields of their activity, opinions on the expected changes in economic environment and conditions of their operation. In the model the business survey data have been used as proxies of firms' expectations concerning the difficulties in production and sales. The excess domestic demand (consumption, investment and imports) has been derived by using the ratio of firms expecting capacity utilization problems. The ratio of firms having serious difficulties in the supply of raw materials and intermediary goods has been incorporated in the production function. The ratio of firms anticipating demand decline has been used in the export functions.

The first version of the model (estimated on the basis of time series for 1970–1987) has been used for ex post simulations covering the eighties in order to evaluate the dynamic properties. The indicators computed for characterizing accuracy of the dynamic simulations are summarized in the table 2.3.

Table 2.3:
The Mean Absolute Percentage Errors (MAPE) and the Root Mean Square Errors (RMSE)

Variables	MAPE (%)	RMSE (%)
Private consumption	0.85	0.99
Total consumption	0.72	0.85
Investments	2.35	2.80
Decentralized	3.37	4.59
Central	2.14	2.43
Exports	1.94	2.12
Rouble	1.93	2.42
Non-roubles	2.32	2.63
Imports	3.30	3.73
Rouble	2.03	2.48
Non-rouble	4.77	5.30
GDP	1.45	1.57

Forecasts first have been prepared for 1990. Using the re-estimated version of the model, medium-term predictions have been computed for 1990–93, in order to check model outcomes for a period within which a shift to a more favorable growth path may be expected. At the beginning of the current year, when model predictions were being prepared, neither the actual data for 1990 nor the figures of the government program had been published. Values of the exogenous variables were based on assumptions.

In the first version of model predictions no significant changes in the income redistribution have been supposed and exogenous variables for the years 1992–93 (concerning price formation, foreign debt increase, tensions in the production) got the same values as in 1991. In this version neither the significant growth decline, nor the drop in investment in 1990 could be forecasted. The results proved that constancy of parameters derived from historical data cannot be hold and highlighted the main inadequacies of the model.

In the second version the relevance of model estimates for the state-owned sector has been investigated. The growth of GDP in 1990 was modified taking into consideration the decrease in the rouble trade and the share of state-owned enterprises' disposable income in the GDP was supposed to be reduced. In this version the new enterprises' investment activity was expected to increase from 1991 and it was given exogenously. For the years 1992–93 slight reduction of inflation and significant improvement in the business survey variables have been supposed. Evidently the forecasts could reflect only the consequences of changes in factors the exogenous variables cover, and could not consider any parameter changes. The results of this second simulation are given in the table 2.4.

Table 2.4:
Forecasts of the Main Macroprocesses for 1990–93
(at 1981 prices, previous year = 100)

	1990*		1991	1992	1993
	A	M			
Gross Domestic Product	95.7	97.0	96.1	104.7	105.1
Private consumption	95.5	96.2	97.0	99.6	100.1
Investments	91.3	97.5	94.1	97.1	99.5
Decentralized		98.2	91.3	94.8	96.9
Central		94.1	92.4	94.6	99.9
Private sector**		100.0	103.0	104.6	105.0
Foreign trade***					
Exports		93.6	99.3	105.7	105.4
Central Eastern Europe	74.0	74.0	82.0	107.5	107.0
Other countries	109.8	107.4	107.7	105.1	104.8
Imports		93.4	91.7	103.3	102.8
Central Eastern Europe	82.2	81.0	75.0	99.5	100.0
Other countries	102.9	103.2	101.6	105.2	103.9

* A: preliminary actual data; M: model forecasts.
** New enterprises.
*** Exports and imports to 'Other countries' here correspond to the former non-rouble trade, and Central and Eastern Europe represents the former rouble trade. Actuals for 1990 are computed without re-exports/imports.

All the results have verified that the shift to a more dynamic growth path may be realized only if the private sector activity and productivity exceed significantly the past performance of the former state-owned sector. Considering our results it may be concluded that in the transition period annual macroeconometric models have been invalidated by the radical regime changes, as the new macro level response parameters cannot be derived without a detailed analysis of the structural changes.

Time series based econometric estimates are less disturbed concerning the households' consumption, where the effects of the accelerating inflation could have been tested yet at the end of the observation period. Besides the income and price effects the influence of wealth may become more and more significant. In Király-Körösi [1990] the negative effect of housing on the long-run equilibrium path of income and consumption has been verified. Although the approach based on time series seems to be convenient for households' consumption, more detailed analysis taking into account the different commodity groups is desirable. Effects of the price liberalization, and changes of the consumption-saving behavior of the population during the high inflation period could be described by disaggregating consumption expenditures. Characteristics of the adjustment process may be refined using quarterly (monthly) data.

The primary outputs of the model are growth rates. One feasible method of producing macroforecasts in the first few years of the transition is that, keeping this way of specification, the forecasted growth rates may be supposed to be valid for the state-owned sector. For the identification of possible parameter and model changes complementary computations should be prepared. The parameters of private sector's model may be determined by 'judgmental techniques', taking into account business surveys and experts' opinions. When constructing macroforecasts the change in weights of the different sectors should be predicted. The privatization process of state-owned enterprises is expected to accelerate, therefore the critical condition of the macroforecasts feasibility is that statistical data concerning privatized enterprises should be separated from the data of the former state-owned sector.

Concerning the supply side, even if sectoral production functions are estimated, the most relevant tendencies cannot be captured on the basis of time series. The crucial point of the growth path shift is that the factor productivity of the new enterprises and that of the privatized firms should be different from that of the state-owned sector, and the increasing competition is expected to result in a more effective operation of the state-owned sector as well. Analysis based on cross sectional data bases (containing representative subsamples of both private and public sectors) may give a chance to characterize this adjustment process.

The fixed capital formation, playing a crucial role in the technological innovation, can be characterized by the same structural problems as production. In addition not only the parameters but the whole 'theoretical' model may be expected to change in the transition period. Investment decisions of the private sector will very likely be based on cost and profitability considerations and instead of variables representing the central control, financial (interest rates, loans) and business cycle indicators should be included in the models. For the decreasing public sector the estimated functions of decentralized and centrally decided investments may be accepted. The private investment activity can be estimated from business surveys.

It is very difficult to define samples for cross sectional computations. In this field a lot of statistical and methodological problems may occur. There are some new problems in statistical data collection for the public sector. Representative samples for state-owned enterprises can be selected for the last years of the eighties but without assuming the homogeneity of the samples (the large state-owned enterprises are very often divided in units before privatization). Problems of sample selection resulting from slowly updated business registers should be solved. The statistical data processing will face a lot of problems due to the introduction of a representative business survey based inference system in macrodata 'production'. The data service for private sector is not yet regular, but there are some occasional business surveys aiming at producing statistics on the base of selected samples (newly established firms, small and medium size enterprises etc.). The change in the composition of samples for both private and public sectors is one of the very important characteristics of transition. It is evident, however, that a lot of methodological difficulties are stemming from the inhomogeneity of samples.

The introduction of convertible currency settlement in foreign trade with Central and Eastern European countries invalidated any forecasting based on time series. At present, on macro level, there is no other way for a macroforecast of trade with these countries than accepting estimations (considering government contracts) made by policy makers. The trade with the other countries may be described and forecasted on the basis of time series but in order to increase the accuracy of forecasts, disaggregation of exports and imports (commodity groups or sectors) is indispensable. For the investigation of the market reorientation time varying models may be tested. Information concerning parameter changes may be derived from micro level analysis, on the basis of cross sectional and panel data. To find the best solution how to incorporate the results of micro level computations to a macrosystem seems to be one of the most important methodological tasks in modeling during the transition period.

A general revision is needed in the macroforecasts concerning the following fields: the price formation, the international indebtedness and the financial processes. The use of short term (quarterly or monthly) data would be indispensable in order to describe dynamic adjustments of these processes.

The hypothesis of exogenous domestic prices should be replaced by a description of a market oriented price system. Besides cost factors (wage rates) and competitive effects between different markets (the foreign market prices that can be maintained exogenous and exchange rate policy) the role of expectations should be investigated.

The foreign indebtedness position could be described in a more detailed way computing some indicators (such like debt to GDP, debt service ratio, reserve ratio etc.) on the basis of which it is possible to forecast the change in the creditworthiness of Hungary.

During the transition period the first steps should be made to make models financially integrated. The part of the income generation and redistribution process, consisting of equations with exogenous parameters, may be replaced by estimated ones as if market conditions are prevailing and the system of new financial regulators is stabilized. For the description of financial behavior of different economic agents, portfolio choice models based on monthly data may be estimated. Because of the shortness of time series (data for about three years are available) the short and longer run effects cannot yet be separated and verified. Institutional and operational changes in the banking system (established in 1987) do not yet show the stability on the basis of which some behavioral equations could be estimated. However a very aggregated flow of funds table can be constructed, which can be used for the characterization of the different agents' financial position, and some point estimations may be prepared concerning the reaction parameters.

5 Conclusions

In this paper the relevance of macroeconometric models in the transition period has been discussed with the conclusion that these models have been invalidated by radical regime change. The date of regime change coincides with the first year of the models' forecasting period, therefore some additional elements are indispensable to track the most important consequences. The elaboration of new, financially integrated econometric models estimated on the basis of short-term time series would be a good solution, but Hungarian short term data supply has not yet been made conform to the annual national accounting system. Establishing a data collection and processing system concerning the private sector is in progress.

Hence, as a provisional solution, macroforecasts in the transition period may be produced only by elaborated systems containing both econometric and 'judgmental' elements. Macroeconometric models based on annual time series, containing the main equilibrium conditions of the economic development, may be considered only as a framework of the forecasting system, that should be enlarged in order to allow of a more complex description of interrelationships between real, financial and monetary processes.

The estimated equations of macromodels may serve as a starting point for characterizing the former state-owned sector's performance, but the expected parameter changes as well as the characteristics of private sector should be investigated on the basis of cross sectional (panel) data. Representing expectations and qualitative information, the role of household and business surveys is getting more and more important in forecasts. The results made by different approaches should be incorporated in the macroeconomic forecasting system, aiming at scenarios of the main processes and determination of the equilibrium position.

APPENDIX

Table 2.A1:
Macro-economic Indicators (Year-to-year rate of change in percent, at constant prices)

	1990	1991	1992	1993*
Gross output	-5.3	-14.1	-7.0	-2–0
of which:				
industry		-21.0	-10.0	0–3
GDP	-3.5	-11.9	-4.5	-2–0
Consumption	-2.7	-5.3	-2.0	. . .
– private	-3.6	-5.8	-2.5	1–2
– public	2.6	-2.7	-4.0	20+
Investments	-7.1	-11.6	-7.0	0–3
Exports	-5.3	-5.0	1.0	-15– -10
Imports	-4.3	5.0	-7.0	5–10
PPI	22.2	28.5	15.5	12.5
CPI	28.9	35.0	23.0	22.5
Terms of trade	0.4	-10.4	-0.5	-1.0
Savings ratio**	5.6	17.1	11.5	8.0
Unemployment rate	1.7	8.5	12.3	12.2
Central budget deficit (BnHUF)	-1.	-114.2	-197.1	-199.7
Percent of GDP	-0.1	-5.0	-7.1	-6.0
Current account (in MnUSD)	127	267	324	-3.4

Source: Main Indicators of Economic Development, 1983-1989, CSO Budapest, National Bank of Hungary Monthly Report 1993/1

* 1993: Preliminary data; Ministry of Finance.

** Households savings in percentage of the disponible income.

\+ Including the effect of amortization of roubles debt by airplanes delivery.

Table 2.A2:
Estimated Equations of the ERI Model[2].

1. Private consumption (QCL) (figure 2.A2)

1/A. Dependent variable: LQCL=Log(QCL)				**1961-1990**
Expl. Variable	Parameter	Standard error	T-ratio	
CONS	.36191	.047586	7.6053	
LRLJOV	.93671	.007560	123.9122	
RJPCL	-.03026	.005831	-5.1892	
D7990	.02941	.004748	6.1928	
R2 = .99937	F(3, 26)	= 13672.4		
R2* = .99929	SE	= .007820		
DW = 1.0809				
AU = CHI-SQ(1)	= 0.2224;	JB = CHI-SQ(2)		= 2.0423
HS = CHI-SQ(1) = 1.4886				
D-F (1962 1990)	= -4.8053(	-4.4962)		
ADF(1)(1963 1990)	= -5.5933(	-4.5113)		
ADF(2)(1964 1990)	= -5.1272(	-4.5276)		
ADF(3)(1965 1990)	= -5.3557(	-4.5452)		
1/B. DQCL=LQCL-LQCL(-1)				**1962-1990**
Expl. Variable	Parameter	Standard error	T-ratio	
CONS	.0052929	.0034880	1.5175	
DRLJOV	.79569	.081654	9.7447	
DDPCL	-.18153	.076605	-2.3697	
RQCL(-1)	-.81315	.22825	-3.5625	
R2 = .88551	F(3, 25)	= 64.4553		
R2* = .87177	SE	= .008750		
DW = 1.8610				
AU = CHI-SQ(1)	= 1.3266;	JB = CHI-SQ(2)		= 3.5581
HS = CHI-SQ(1)	= 2.6477;			

DPCL – inflation rate (CPI); DDPCL=DPCL-DPCL(-1);
RLJOV – real disposable income of households;
LRLJOV = Log(RLJOV);
RJPCL = LRLJOV*DPCL;
RQCL – are residuals from 1./A equation.
DU7990 – equals to 1 for 1979-1990, 0 otherwise.

[2] Variables at constant and those at current prices are marked by symbols beginning with 'Q' and 'X', respectively. The logarithm of a given variable is denoted by 'L' and differences of logs are written as 'D'. R^2 stands for the adjusted multiple correlation coefficient and DW, AU, JB, HS, D-F, ADF denotes the Durbin-Whatson, LM of residual serial correlation, normality and heteroscedasticity, Dickey-Fuller and Augmented DF test statistics for residuals resp.

2. Decentralized investments (QIE) (figure 2.A3, Investments total)

Dependent variable: $DQIE=\log(QIE)-\log(QIE_1)$	**1970-1990**

$$DQIE = \underset{(2.09)}{0.0338} + \underset{(1.38)}{0.1097}\, DQRE - \underset{(-1.35)}{0.1171}\, DBD_1 + \underset{(1.94)}{0.006} \log(1-CAP) -$$

$$\underset{(-5.88)}{-0.7973} [LQIE_1 - 0.4985\, LQRE_1 - 0.3476\, LQRBA_1]$$

$$R^2 = 0.7020 \qquad DW = 1.8557$$

where

QRE – real disposable revenue of enterprises; nominal revenue per investment price index (QRE = RE/PI);

QRBA – budget subsidies for investments deflated by investment price index (QRBA=RB /PI);

BD – growth rate of the convertible gross debt.

CAP – business survey data: the ratio of enterprises declaring capacity utilization problem.

3. Centrally decided investments (QIA) (figure 2.A3)

Dependent variable: $DQIA=\log(QIA)-\log(QIA_1)$	**1970-1990**

$$DQIA = \underset{(1.65)}{0.4499} \underset{(-1.77)}{-0.1129}\, DQIA_1 + \underset{(3.3)}{0.2135}\, DU87 + \underset{(3.66)}{0.0082} \log(1-CAP)$$

$$+(\underset{(1.44)}{0.1145} - \underset{(-1.57)}{0.0417}\, T)\, LBD_1$$

$$R^2 = 0.6797 \qquad DW = 2.3825$$

where – DU87 – dummy: equals to 1 in 1987.

4. Rouble exports (QER) (figure 2.A4, Exports totale)

Dependent variable: DQER=log(QER)-log(QER_1)	**1970-1990**

$$DQER = \underset{(-2.04)}{-0.9115} + \underset{(2.32)}{0.1926}\, LRPER \; \underset{(-2.48)}{-0.1465}\, \log(1-ED) +$$

$$+ \underset{(2.81)}{0.0133}\log(1-CAP) \; \underset{(-3.14)}{-0.2534}\, [LQER_1 - \underset{(-13.24)}{1.3224}\, LQMR_1]$$

$$R^2 = 0.6079 \qquad DW = 2.2708$$

where

- QMR – rouble imports;
- RPER – ratio of the modified rouble price index (PER + net export subsidy coefficient, SER), to that of the non-rouble exports (PED+SED);
- ED – business survey data: the ratio of firms having non-rouble market problems.

5. Non-rouble exports (QED)

Dependent variable: LQED=log(QED)	**1970-1990**

$$DQED = \underset{(-1.85)}{-0.2975} + \underset{(2.07)}{0.0750}\, LRPED_1 + \underset{(2.44)}{0.0694}\, LR1PED_1 + \underset{(1.95)}{0.1968}\, DKONJ$$

$$- \underset{(-2.95)}{0.5181}\, [LQED_1) - 0.5290\, LKONJ]$$

$$R^2 = 0.5159 \qquad DW = 2.3068$$

where

- RPED – ratio of domestic price of exports to price index KOPINT, RPED=(PED+SED)/KOP[3]; from 1983 equals to 0.
- R1PED – ratio of rouble and non-rouble export price indices corrected by subsidies, from 1983;
- KONJ – index of export demand: computed as the weighted average of main partners' import growth index, where weights are the shares of the given partner in Hungarian exports;

[3]KOP represents the world market price index published by the Hungarian KOPINT Research Institute. This index combines world price changes which are important from the point of view of Hungarian foreign trade.

6. Rouble imports (QMR) (figure 2.A4, Imports totale)

Dependent variable: LQMR=log(QMR)	**1960-1990**

$$LQMR = \underset{(-17.68)}{-5.7699} + \underset{(32.58)}{1.6510}\, LDDEM^*$$

$$R^2 = 0.9815 \qquad DW = 2.1236$$

where

DDEM* – domestic demand (Public expenditures QCK + notional consumption and investment demand QCP* + QCI* computed from consumption and investment equations taking into account the effect of business survey information.)

7. Non-rouble imports (QMD)

Dependent variable: LQMD=log(QMD)	**1960-1990**

$$LQMD = \underset{(-4.30)}{-4.4367} + \underset{(7.38)}{1.8811}\log(QCP^*+QI^*) + \underset{(2.55)}{0.0364}\log(1-CAP)$$

$$+\underset{(1.46)}{0.2112}\, LTB_1 \underset{(-1.79)}{-0.6245}\, LRPMS$$

$$R^2 = 0.9171 \qquad DW = 1.7004$$

where

QMD – non-rouble imports;

TB – non-rouble trade balance at current domestic price (TB = QED PED -QMD PMD);

RPMS – ratio of the modified non-rouble import price index (PMD+SMD, SMD is the net import taxes coefficient) to domestic producer prices (PX).

8. Gross Domestic Product (QGDP) (figure 2.A1)

Dependent variable: RQGDP=Log(QGDP/L^+)	1970-1990

$$RQGDP = \underset{(-4.74)}{-2.1594} + \underset{(1.92)}{0.0966} \log(CU^* K^+/L^+) + \underset{(10.64)}{0.0362}\, T$$

$$+ \underset{(5.48)}{0.4619}\,(1- DI)$$

$R^2 = 0.9911$ DW = 1.4557

where

- L^+ – average number of active wage-earners;
- K^+ – average stock of gross fixed capital;
- CU – capacity utilization;
- DI – business survey data: ratio of enterprises declaring difficulties in raw material and intermediary good supply.

Figure 2.A1:

Gross Domestic Product
(previous year = 100, at 1981 prices)

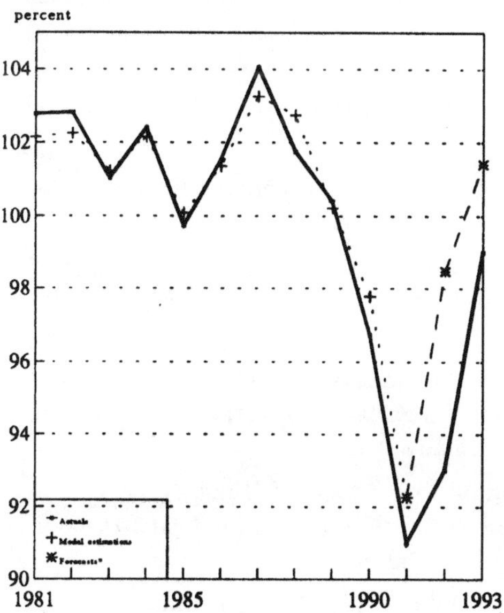

Figure 2.A2:

Investment
(previous year = 100, at 1981 prices)

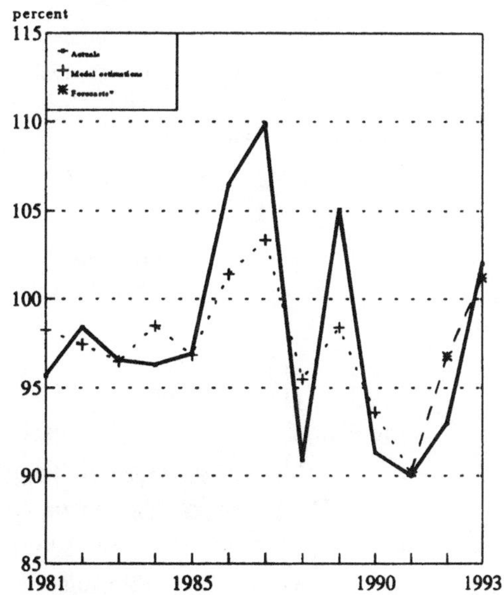

Figure 2.A3:

Private Consumption
(previous year = 100, at 1981 prices)

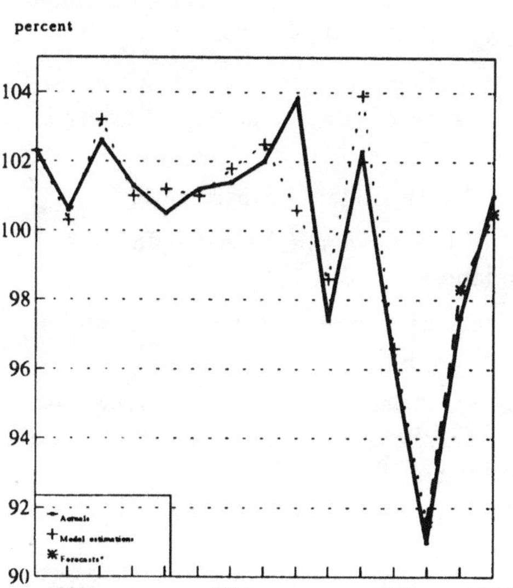

Figure 2.A4:

Exports and Imports
(previous year = 100, at 1981 prices)

* The model was estimated on the basis of 1970-90 data.

References

Halabuk, L., K. Hulyák, Zs. Nyáry, and M. Kotász (1973): "A magyar népgazdaság ökonometriai modellje (Econometric model of Hungary)." Akadémiai Könyvkiadó.

Halabuk, L. and H. T. Shapiro (1976): "Macroeconometric Model Building in Socialist and non-socialist Countries: a Comparative Study." *International Economic Review*, **66**: 871–879.

Halpern, L. (1989): "Effects of Devaluation in a Macro-econometric Model for Hungary." *Acta Oeconomica*, No. 3-4, pp. 293–312.

Hulyák, K. (1982): "Disequilibrium Models for Consumption Goods in Hungary." Paper presented to International Conference of Applied Econometrics, Budapest.

Hulyák, K. (1986): "Jövedelem és árhatások a magyar dollárexportban és importban (Income and Price Effects in Hungarian Convertible Exports and Imports)." Tervgazdasági Intézet Tanulmányok.

Hulyák, K. (1989): "Macroeconomic Disequilibrium Model of Hungary." In: C. Davis and W. Charemza, Chapman and Hall (eds.), Models of Disequilibrium and Shortage in the Centrally Planned Economies.

Hunyadi, L., J. Neményi, P. Subicz, and A. Fiala (1979): "A rövidtávú tervezés ökonometriai modellje (Econometric Model for the Short-Term Planning)." Közgazdasági és Jogi Könyvkiadó, Budapest.

Hunyadi, L., Å. Kovács, and J. Neményi (1985): "An Econometric Input-Output Model of Hungarian Economy." Paper presented at the conference on 'Econometric Modelling of the Socialist Economies', Bratislava.

Király, J. and G. Kôrösi (1990): "Consumption and Housing in Hungary." Paper presented at the Sixth World Congress of the Econometric Society in Barcelona.

Kornai, J. (1980): "Economics of Shortage." North-Holland, Amsterdam.

Neményi, J. (1990): "Annual Macromodel of the Economic Research Institute." ERI Working Papers August 1990, Budapest.

Portes, R. (1984): "The Theory and Measurement of Macro-economic Disequilibrium in CPEs." Paper No. 91, CEPR, London.

Simon, A. (1979): "A termelés és a külkereskedelem kapcsola-tának vizsgálata ökonometriai módszerrel Econometric Analysis of the Foreign Trade and the Output)." Konjunktúra ás Piackutató Intézet, Budapest.

Wallis, K. F. (1989): "Macroeconomic Forecasting: a Survey." *The Economic Journal*, **99**: 28–61.

Chapter 3

Pooling Noisy Data Sets

EDWARD E. LEAMER*

1 Introduction

Empirical studies of the Eastern European economies for the foreseeable future will have to make due with data sets that are limited in terms of quantity, quality and relevance. Facing this scarcity of useful data, analysts will have to import data from analogous Western countries. Many of these imports will be buried in the baggage of ideas that Western economists have formed from observation of Western economies. These hidden stowaway ideas will make it difficult to evaluate the inferences that Western economists draw from Eastern data. We will be forced to guess if an analyst is acting as if Hungary is more analogous to Austria or more analogous to Spain. It may be better to pool Western with Eastern data in a formal econometric way, and thereby to develop a language that facilitates the conversation about the strength and the importance of the analogies on which the pooling depends.

This paper proposes an econometric method of pooling Western and Eastern data for the estimation of a linear regression model. The method is Bayesian and uses prior information about the regression coefficients. The pooled estimates depend on three parameters: (1) δ, the lack of confidence in the prior information, (2) ρ, the degree of similarity of Western and Eastern economic structures, and (3) λ, the amount of experimental contamination

*Support from NSF grant SES 8910950 is gratefully acknowledged. Able computational assistance was provided by Rodrigo Fuentes. The data set has been provided by Robert Barro of Harvard University and the NBER and is available on request.

caused by low-quality and/or irrelevant data, left-out variables, simultaneity and other pathologies.

Assumptions about these three items will be necessary whenever data sets are pooled whether formally or informally. The formal treatment suggested in this paper is simple and direct. It makes obvious how the pooling depends on these three assumptions. The simplicity and transparency of this framework should facilitate the conversation that ought to occur about the role that Western and Eastern analogies play in Eastern policymaking.

But a formal approach is not without serious shortcomings. One shortcoming is that it requires more work. More importantly, mathematical, numerical and cognitive limitations force great simplifications and a high degree of inflexibility. Though pooling of data sets will necessarily depend on the confidence in prior information, the degree of similarity of the structures and the amount of experimental contamination, these vague concepts have been translated into a precise mathematical model for inference in a way that may leave you so uncomfortable that the mathematics impedes the conversation. Ultimately it is up to the consumer to decide if the cost in terms of inflexibility is worth the increase in clarity.

The principal methodological contribution of this paper is the addition of the experimental bias parameters into the pooling problem. The traditional econometric methods of pooling data sets are intended to deal with data that are limited in terms of quantity, but not in terms of quality and relevance. The experimental bias parameters allow the data sets to be of doubtful relevance and quality.

The traditional methods for pooling data sets use the random coefficients model in which parameters applying to different experiments are treated as if they were a random sample out of a population. Methods such as "the random coefficients model", "Type II Analysis of Variance", "Bayes" or "empirical Bayes" differ primarily in how they treat uncertainty about the hypothetical population of parameters. For example, Aigner and Leamer [1984], use an empirical Bayes approach to pool time-of-use pricing experiments that were performed by different electric utilities. DeMouchel and Harris [1981] use Bayes and empirical Bayes approaches to "combine cancer experiments in man and other species".

These approaches generally presume that the data are perfect in quality and relevance. Another tradition deals with dubious data by referring to errors in measurement in the variables. Measurement error causes bias in the estimates of regression coefficients, which can be corrected only given some information about the probable amount of the measurement errors. Absent that additional information, the model is underidentified and a data set admits a set of equally good estimates – the so-called errors-in-variables bound. See for example, Leamer [1987], Klepper and Leamer [1984], Aigner et al. [1984] and Patefield [1981].

In a companion paper, Leamer [1991] directly combines these two statistical traditions to deal with data sets that are both brief and noisy, but with identical structures and no prior information. The errors-in-variable model is complex to begin with and when it is combined with the random coefficients model which would allow differences in structures, the analysis becomes quite difficult. In this paper, the errors-in-variable model is reparameterized to allow a somewhat easier treatment. The explanatory variables are entered twice in the regression model, once to represent the "true" or "structural" coefficients and a second time to represent the experimental bias associated with measurement errors, and other statistical problems that cause bias. This "contaminated" regression model has been used by Leamer [1974] and Leamer [1978] to discuss data-instigated models.

Traditionally, errors-in-variables issues would be analyzed with a measurement model and prior information about the bias would be entered indirectly through prior information about the signal to noise ratio in the measurement model. Here the path is the opposite direction. Prior information about the signal to noise ratio is being introduced implicitly through prior information about the bias parameters. The choice of approach depends on economy and accuracy. It is clearly more economical to use the bias parameters rather than the signal-to-noise ratios since the econometric procedures that refer to bias parameters are an order of magnitude easier than procedures that refer to a measurement model. But accuracy is another matter. If your prior information refers directly to signal-to-noise ratios, you may find it very difficult to think about the bias parameters which are complicated functions of the signal and noise covariance matrices.

The symbols that represent the three critical inputs are (1) δ, the doubt about the prior (2) ρ, the structural similarity, and (3) λ, the experimental bias. The traditional pooling model uses the data to estimate κ, the degree of similarity in the structural parameters, but takes special values of δ and $\lambda_i : \delta = \infty$ (No prior information), and $\lambda_i = 0$ (No experimental bias).

Doubt about the values of these "hyperparameters" can be treated either with a sensitivity analysis or with an estimation approach. The discussion that follows presumes that there are only two data sets to be pooled[1]. With a

[1]If there are more than two data sets, additional contamination parameters, $\tilde{N}$, would be required. Furthermore, a single correlation parameter ρ indicating the similarity of

sample size of only two it is impossible with accuracy to infer much about the distribution from which the regression parameters are drawn and a sensitivity analysis may be the preferred approach. Both Bayesian estimation of these hyperparameters and also sensitivity analysis are proposed.

The method is illustrated by a study of the determinants of the growth rates of developed and developing countries. These two data subsets yield estimates that are very similar, which makes the results not very sensitive to the form of pooling and which points toward low experimental bias and a high degree of similarity.

2 Bayesian Pooling of Contaminated Data Sets

A Bayesian approach for the pooling of contaminated data sets makes use of the contaminated regression model:

$$\boldsymbol{y}_i = \boldsymbol{X}_i\boldsymbol{\beta}_i + \boldsymbol{X}_i\boldsymbol{\theta}_i + \boldsymbol{\varepsilon}_i \qquad i = 1,2$$
$$\boldsymbol{\varepsilon}_i \sim \boldsymbol{N}(0, \sigma_i^2 \boldsymbol{I})$$

where $\mathbf{y}_i$ is an $n_i \times 1$ vector of observations of the "dependent" variable, $\mathbf{X}_i$ is an $n_1 \times k$ matrix of observations of the k "explanatory" variables, $\boldsymbol{\beta}$ is the $k \times 1$ vector of "true" parameters and $\boldsymbol{\theta}$ is the $k \times 1$ bias vector representing the experimental contamination due to measurement errors or any other statistical pathologies. This model suffers from an extreme multicollinearity problem: all the variables enter twice, once to capture the structural effects $\boldsymbol{\beta}$ and again to capture the experimental contamination $\boldsymbol{\theta}$. The informational deficiencies of this underidentified model can be overcome by supplementing the data information with prior information. The prior that seems natural should embody the ideas that the experimental contamination is probably small, and that the structural parameters are probably similar in the two data sets. Smallness of the experimental contamination can be captured with a prior located at the origin:

$$\boldsymbol{\theta}_i \sim \boldsymbol{N}(0, \boldsymbol{V}_i),$$

where $\boldsymbol{V}_i$ is the prior covariance matrix, the smaller its value the smaller is the probable experimental contamination. A prior for the structural parameters

structures is unlikely to be judged adequate. A variety of higher dimensional correlation matrices might be used but one that has some appeal begins with data sets ordered by their similarity and assigns a correlation depending on the differences in their orders. (Like a stationary time series.)

that captures information about their probable sizes as well as the idea that they are similar is the normal distribution

$$\begin{bmatrix} \boldsymbol{\beta}_1 \\ \boldsymbol{\beta}_2 \end{bmatrix} \approx \boldsymbol{N}\left(\begin{bmatrix} \boldsymbol{p} \\ \boldsymbol{p} \end{bmatrix}, \begin{bmatrix} \boldsymbol{U} & \rho\boldsymbol{U} \\ \rho\boldsymbol{U} & \boldsymbol{U} \end{bmatrix} \right)$$

Here the parameter ρ is the correlation of the structural parameters across the two data sets, the vectors $\boldsymbol{p}$ represent the most likely values of these structural parameters and the covariance matrix $\boldsymbol{U}$ measures the likely departures from $\boldsymbol{p}$. Incidentally, this parameterization allows a relative lack of information about the values of $\boldsymbol{\beta}$ but confidence that $\boldsymbol{\beta}_1 - \boldsymbol{\beta}_2$ is small. This can be accomplished by selecting a large value of $\boldsymbol{U}$ and a value of ρ close to one such that $Var(\boldsymbol{\beta}_1 - \boldsymbol{\beta}_2) = 2\boldsymbol{U}(1-\rho)$ is small.

With these as the elements, the full prior covariance matrix takes the form

$$V = Var\left(\begin{bmatrix} \boldsymbol{\beta}_1 \\ \boldsymbol{\beta}_2 \\ \boldsymbol{\theta}_1 \\ \boldsymbol{\theta}_2 \end{bmatrix} \right) = \begin{bmatrix} \boldsymbol{U} & , \rho\boldsymbol{U} & , 0 & , 0 \\ \rho\boldsymbol{U} & , \boldsymbol{U} & , 0 & , 0 \\ 0 & , 0 & , \boldsymbol{V}_1 & , 0 \\ 0 & , 0 & , 0 & , \boldsymbol{V}_2 \end{bmatrix}$$

The corresponding sample precision matrix is

$$N = \begin{bmatrix} \boldsymbol{X}_1'\boldsymbol{X}_1\sigma^2 & , & 0 & , & \boldsymbol{X}_1'\boldsymbol{X}_1\sigma^2 & , & 0 \\ 0 & , & \boldsymbol{X}_2'\boldsymbol{X}_1\sigma^2 & , & 0, & , & \boldsymbol{X}_2'\boldsymbol{X}_1\sigma^2 \\ \boldsymbol{X}_1'\boldsymbol{X}_1\sigma^2 & , & 0 & , & \boldsymbol{X}_1'\boldsymbol{X}_1\sigma^2 & , & 0 \\ 0 & , & \boldsymbol{X}_2'\boldsymbol{X}_1\sigma^2 & , & 0 & , & \boldsymbol{X}_2'\boldsymbol{X}_1\sigma^2 \end{bmatrix}$$

and the sample cross-product matrix is

$$r = \begin{bmatrix} \boldsymbol{X}_1'\boldsymbol{y}_1/\sigma_1^2 \\ \boldsymbol{X}_2'\boldsymbol{y}_2/\sigma_2^2 \\ \boldsymbol{X}_1'\boldsymbol{y}_1/\sigma_1^2 \\ \boldsymbol{X}_2'\boldsymbol{y}_2/\sigma_2^2 \end{bmatrix}$$

If we denote the stacked vector of parameters by γ:

$$\gamma = \begin{bmatrix} \boldsymbol{\beta}_1 \\ \boldsymbol{\beta}_2 \\ \boldsymbol{\theta}_1 \\ \boldsymbol{\theta}_2 \end{bmatrix} \qquad E(\gamma) = \begin{bmatrix} \boldsymbol{p} \\ \boldsymbol{p} \\ 0 \\ 0 \end{bmatrix}$$

the Bayesian posterior moments are (e.g. Leamer [1978] p. 78):

$$E(\gamma|\boldsymbol{y}_1, \boldsymbol{X}_1, \boldsymbol{y}_2, \boldsymbol{X}_2, \sigma_1^2, \sigma_2^2) = (\boldsymbol{N} + \boldsymbol{V}^{-1})^{-1}(\boldsymbol{r} + \boldsymbol{V}^{-1}E(\gamma)) \tag{3.1}$$

$$Var(\gamma|\boldsymbol{y}_1, \boldsymbol{X}_1, \boldsymbol{y}_2, \boldsymbol{X}_2, \sigma_1^2, \sigma_2^2) = (\boldsymbol{N} + \boldsymbol{V}^{-1})^{-1} \tag{3.2}$$

These two posterior moments condition on the residual variances σ_1^2 and σ_2^2, which rarely would be known. A prior distribution for these parameters is required to deal with the uncertain case, but, unfortunately, there is no prior distribution that leads to a tractable nonapproximate posterior distribution. A traditional prior distribution would be the product of gamma distributions on σ_1^2 and σ_2^2. Then after integrating these parameters from the likelihood function, the posterior is the product of a Student distribution for $(\boldsymbol{\beta}_1, \boldsymbol{\theta}_1)$ times a Student distribution for $(\boldsymbol{\beta}_2, \boldsymbol{\theta}_2)$ times the normal prior for $(\boldsymbol{\beta}_1, \boldsymbol{\theta}_1, \boldsymbol{\beta}_2, \boldsymbol{\theta}_2)$. This complicated density can have many modes but will look like a unimodal normal distribution if the two Student pieces are approximately normal. This occurs when the data and prior combine to firmly establish the values of σ_1^2 and σ_2^2 and we then revert approximately to the former case of known σ_1^2 and σ_2^2. In this paper, I will act as if the values of σ_1^2 and σ_2^2 can be accurately estimated from the data and treated as known. The estimates that I will use are:

$$\sigma_i^2 = \frac{\boldsymbol{y}_i(\boldsymbol{I} - \boldsymbol{X}_i(\boldsymbol{X}_i'\boldsymbol{X}_i)^{-1}\boldsymbol{X}_i')\boldsymbol{y}_i}{n_i - k} \tag{3.3}$$

where n_i is the dimension of $\boldsymbol{y}_i$ and k is the dimension of $\boldsymbol{\beta}$. An improved approximation would make use of the prior information about the slope parameters and use the (iterative) estimate

$$\sigma_i^2 = \frac{(\boldsymbol{y}_i' - \boldsymbol{X}_i\boldsymbol{\beta}_i)'(\boldsymbol{y}_i' - \boldsymbol{X}_i\boldsymbol{\beta}_i)}{n_i} \tag{3.4}$$

where $\boldsymbol{\beta}_i$ is evaluated at, or at least near, the posterior mode. The approximation (3.3) will be accurate if (3.3) and (3.4) are not very different and if the degrees of freedom $n_i - k$ are both adequately large.

2.1 Sensitivity Analysis

It is highly unlikely that the prior parameters $(\boldsymbol{U}, \boldsymbol{\rho}, \boldsymbol{V}_1, \boldsymbol{V}_2, \boldsymbol{p})$ could be chosen with complete confidence. If the data are not sufficiently informative about these parameters a sensitivity analysis is required to identify those inferences that are too sensitive to the choice of prior to be taken seriously. The sensitivity analysis is facilitated if the number of prior parameters is reduced. A natural constraint is

$$\boldsymbol{V}_i = \lambda_i \boldsymbol{U},$$

where λ_i measures the relative importance of experimental contamination. Furthermore, since it will be difficult to carry on a sensitivity analysis for all elements of the prior covariance matrix $\boldsymbol{U}$, it seems better only to perturb $\boldsymbol{U}$ by a scalar $\boldsymbol{\delta}$:

$$\boldsymbol{U} = \boldsymbol{\delta}^2 \boldsymbol{U}_0$$

Thus for this sensitivity analysis the initial prior covariance matrix $\boldsymbol{U}_0$ and the vector of prior means $\boldsymbol{p}$ are taken as given and the scalars $\lambda_1, \lambda_2, \delta$ and ρ are perturbed.

To decide if an inference is excessively sensitive to the choice of these parameters we need to select an sensible amount of perturbation which requires clear thinking about the meaning of each parameter. The parameter δ is a discount rate applicable to the prior information. Formally, it is the factor by which the prior standard errors are multiplied. A value of $\delta = 1$ selects the initial prior covariance matrix. A value of $\delta = 2$ selects a prior covariance matrix with standard errors multiplied by two. A range of $1/4 \leq \delta \leq 4$ seems like a large range.

The parameter ρ measures the similarity of the structural parameters across data sets. Though this correlation could be negative, it is unlikely that there would be many settings in which that would be a sensible choice. For that matter, the reason for pooling must be that the structures are adequately similar and values of ρ in excess of .5 see, seem sensible.

To think about the choice of λ_i it is instructive to consider the problem of a single contaminated data set. Then the posterior mean is a weighted average of the prior mean $\boldsymbol{p}$ and the OLS estimate with weights that depend on λ_i. In the next paragraph it is shown that, as the sample size increases, the weights on prior and sample converge respectively to $\lambda_i/(1+\lambda_i)$ and $1/(1+\lambda_i)$. Thus λ_i measures the resistance to the new information: the larger is the value of λ_i, the less weight is put on the sample result asymptotically. The traditional value is $\lambda_i = 0$, meaning that in a sufficiently large sample, the prior is altogether discarded. A value of λ_i equal to one selects a degree of experimental error that asymptotically assigns half the weight to the prior and half to the data.

To show the influence of λ_i as the sample size grows, we may explore the contaminated model with a single data set. Leamer ([1978], pp. 295–299) reports the following posterior mean:

$$E(\boldsymbol{\beta}_1 | \boldsymbol{y}_1, \boldsymbol{X}_1) = \boldsymbol{D}^{-1}(\boldsymbol{U}^{-1}\boldsymbol{p} + [\boldsymbol{N}_1 - \boldsymbol{N}_1(\boldsymbol{N}_1 + \boldsymbol{\lambda}_1^{-1}\boldsymbol{U}^{-1})^{-1}\boldsymbol{N}_1]\boldsymbol{b}_1)$$

where $\boldsymbol{b}_1$ is the OLS estimate $(\boldsymbol{X}_1'\boldsymbol{X}_1)^{-1}\boldsymbol{X}_1'\boldsymbol{y}_1$, $\boldsymbol{N}_1 = \boldsymbol{X}_1'\boldsymbol{X}_1/\sigma_1$, and $\boldsymbol{D} = \boldsymbol{U}^{-1} + \boldsymbol{N}_1 - \boldsymbol{N}_1(\boldsymbol{N}_1 + \lambda_1^{-1}\boldsymbol{U}^{-1})^{-1}\boldsymbol{N}_1$ where $\boldsymbol{N}_1 = \boldsymbol{X}_1'\boldsymbol{X}_1/\sigma_1^2$. In words, the posterior mean is a weighted average of the prior mean and the sample estimate. If there were no contamination, that is if $\lambda_1 = 0$, then the weight on the prior would be the prior precision $\boldsymbol{U}^{-1}$ and the weight on the data would be the data precision $\boldsymbol{N}_1$. The effect of the sample contamination is to reduce the sample weight to $[\boldsymbol{N}_1 - \boldsymbol{N}_1(\boldsymbol{N}_1 + \lambda_1^{-1}\boldsymbol{U}^{-1})^{-1}\boldsymbol{N}_1] = \boldsymbol{N}_1(\boldsymbol{N}_1 + \lambda_1^{-1}\boldsymbol{U}^{-1})^{-1}$ $(\boldsymbol{N}_1 + \lambda_1^{-1}\boldsymbol{U}^{-1} - \boldsymbol{N}_1) = (\boldsymbol{I} + \lambda_1^{-1}\boldsymbol{U}^{-1}\boldsymbol{N}_1^{-1})^{-1}\lambda_1^{-1}\boldsymbol{U}^{-1}$. Thus as the sample size grows in the sense that $\lambda_1^{-1}\boldsymbol{U}^{-1}\boldsymbol{N}_1^{-1}$ converges to the zero matrix, the weight on the sample converges to $\lambda_1^{-1}\boldsymbol{U}^{-1}$, compared with the sample-independent weight on the prior of $\boldsymbol{U}^{-1}$.

2.2 Estimation

The prior parameters $\lambda_1, \lambda_2, \delta$ and ρ can be perturbed to see how much they matter. They can also be estimated since there is at least some sample information about their values. For example, if the two ordinary least squares estimates are both very precise and very different, the data are incompatible with the assumption that the parameters are very similar ($\rho = 0$) and without experimental error ($\lambda_i = 0$).

The posterior distribution for a parameter is proportional to the product of the prior times the marginal likelihood

$$f(\boldsymbol{y}_1, \boldsymbol{y}_2 | \boldsymbol{X}_1, \boldsymbol{X}_2, \lambda_1\lambda_2\delta, \rho)$$

which is the joint marginal density for $\boldsymbol{y}_1$, $\boldsymbol{y}_2$ evaluated at the sample values. The label "marginal likelihood" refers to the fact that the parameters $\boldsymbol{\beta}$ and $\boldsymbol{\theta}$ have been integrated from the likelihood function. This integration is rather easily accomplished. We may stack the model to form:

$$\begin{bmatrix} \boldsymbol{y}_1/\sigma_1 \\ \boldsymbol{y}_2/\sigma_2 \end{bmatrix} \begin{bmatrix} \boldsymbol{X}_1/\sigma_1 & , & 0 \\ 0 & , & \boldsymbol{X}_2/\sigma_2 \end{bmatrix} \begin{bmatrix} \boldsymbol{\beta}_1 \\ \boldsymbol{\beta}_2 \end{bmatrix} + \begin{bmatrix} \boldsymbol{\varepsilon}_1/\sigma_1 \\ \boldsymbol{\varepsilon}_2/\sigma_2 \end{bmatrix}$$

which in obvious notation can be written as:

$$\boldsymbol{Y} = \boldsymbol{X}\gamma^* + E$$

Then given the normal prior distribution for the vector γ^*, the marginal distribution for $\boldsymbol{Y}$ is normal with mean and covariance

$$E(\boldsymbol{Y}) = \boldsymbol{X}E(\gamma^*), \qquad Var(\boldsymbol{Y}) = \boldsymbol{X} \qquad Var(\gamma^*)\boldsymbol{X}' + \boldsymbol{I}.$$

The corresponding marginal likelihood is therefore:

$$f(\boldsymbol{Y}|\boldsymbol{X}, \lambda_1, \lambda_2, \delta, \rho) \propto |Var(\boldsymbol{Y})|^{1/2} \quad (3.5)$$
$$\exp\{-(\boldsymbol{Y} - \boldsymbol{X} \quad E(\boldsymbol{y}^*))'[Var(\boldsymbol{Y})]^{-1}(\boldsymbol{Y} - \boldsymbol{X} \quad E(\boldsymbol{y}^*))/2\}$$

where the dependence of this statistic on the parameters λ_1, λ_2, δ, and ρ is implicit ($Var(\boldsymbol{Y})$ depends on them.).

This marginal likelihood involves the inverse and determinant of a high-dimensional matrix, a calculation which stretches the computing capacity of most machines. The dimensionality of the problem can be reduced using the following matrix results:

$$|Var(\boldsymbol{Y})| = |\boldsymbol{X} \quad Var(\boldsymbol{\gamma}^*)\boldsymbol{X}' + \boldsymbol{I}| = |Var(\boldsymbol{\gamma}^*)| \quad |[Var(\boldsymbol{\gamma}^*)]^{-1} + \boldsymbol{X}'\boldsymbol{X}|$$
$$[Var(\boldsymbol{Y})]^{-1} = \boldsymbol{I} - \boldsymbol{X}([Var(\boldsymbol{\gamma}^*)]^{-1} + \boldsymbol{X}'\boldsymbol{X})^{-1}\boldsymbol{X}'$$

Using these formulae one needs to compute inverses and determinants of matrices of order $2k$ where k is the number of variables. This contrasts with $Var(\boldsymbol{Y})$ which is of order $n_1 + n_2$ where n_i is the number of observations of type i. Typically $n_1 + n_2$ is larger number than $2k$.

The covariance matrix of the parameters can be expressed in Kronecker notation as $Var(\boldsymbol{\gamma}^*) = (\boldsymbol{A} \otimes \boldsymbol{U}_0)\delta^2$, where

$$\boldsymbol{A} = \begin{bmatrix} 1 + \lambda_1 & , & \rho \\ \rho, & , & 1 + \lambda_2 \end{bmatrix}$$

With $\boldsymbol{A}$ a 2×2 matrix and $\boldsymbol{U}_0$ a 7×7 matrix, results on Kronecker products imply

$$|Var(\boldsymbol{\gamma})| = |\boldsymbol{A} \otimes \boldsymbol{U}_0\delta^2| = |\boldsymbol{A}|^7|\boldsymbol{U}_0|^2\delta^{28} = \left[(1 - \rho^2)\lambda_1\lambda_2\right]^7 |\boldsymbol{U}_0|^2\delta^{28}$$
$$[Var(\boldsymbol{\gamma})]^{-1} = \boldsymbol{A}^{-1} \otimes \boldsymbol{U}_0^{-1}\delta^{-2}$$

3 An Example: The Convergence Hypothesis

In a companion paper, Leamer [1991] has studied the determinants of the growth rate of per capita GNP using a cross-section of countries as has been done by Barro [1991]. That paper deals with the incompatibility of the developed country and developing country subsamples using an errors-in-variables approach but with the assumption that the regression coefficients are the same in both subsamples. Here we combine errors-in-variables concerns with doubt about the similarity of the coefficients. Doubt about the similarity of coefficients is expressed with a random coefficients model that allows variability of regression parameters across observational units. Errors-in-variables concerns enter through a bias parameter vector.

Table 3.1 contains a list of variables together with various summary statistics. The dependent variable is the average annual growth rate of real per capita gross domestic product from 1960 to 1985. The 98 countries that form the data set grew 2.2 percent per annum on the average. The lowest per capita growth rate was minus 1.7 percent; the highest was 7.4 percent. Seven variables are hypothesized to determine the growth rate. A variable of special interest is the initial per capita GDP which is the focus of the convergence literature. A negative coefficient on initial GDP would indicate that countries which are initially ahead tend to grow more slowly; thus there is convergence of per capita GDP. Barro chooses to control for a number of other effects including school enrollment rates, government expenditures, frequency of revolutions and coups, frequency of assassinations and last a measure of macroeconomic disequilibrium – the deviation of investment prices from their mean.

Table 3.2 contains the prior parameters and the OLS estimates of this model. The prior means and standard errors in the first column were chosen to approximate the author's state of mind prior to analyzing this data set. The coefficient of -0.01 on GDP60 means that it would take a change of initial per capita GDP by 1($ 1000) to change the growth rate by .01 (1 percent). Similarly, it would take a 10 percent change in the secondary school enrollment rate, or a 20 percent change in the primary school enrollment rate, or a 100 percent in the share of government, or .1 more revolutions per year or 1 more assassinations per year to alter the growth rate by .01 (1 percent). As I think of this, I am revealing an attitude that the measured share of government probably doesn't matter very much but that political upheavals do matter a lot. Prior standard errors are all set equal to the prior means to suggest a fair degree of confidence in the sign of the coefficients.

The pooled OLS estimates are smaller than the prior means, but all the same sign. The exception is the coefficient on government, which is more than ten times as large as the prior. The other big differences are the coefficients on revolutions and assassinations which are a very small fraction of the prior means. Thus the data seem to want to make me change my mind: government matters more than I thought; revolutions and assassinations don't matter as much.

The regressions on subsets of the data are remarkably similar. The developed country estimates are almost all slightly smaller than the developing country estimates. One exception is the coefficient on government expenditures which is much smaller for the developed than the developing countries. Perusal of table 3.2 thus suggests one puzzle: what is the size of the government effect? Are there assumptions that could justify an estimate as big as the pooled OLS of -0.13 or the even greater developing-country OLS of -0.16? Why is the estimate for the developed countries so much smaller (-0.04)?

Various Bayes estimates are reported in tables 3.3, 3.4, and 3.5. These three tables indicate respectively how the Bayes estimates depend on the three critical inputs:

(1) Inaccuracy of the prior information (δ);

(2) experimental contamination (λ_1 and λ_2); and

(3) structural similarity (ρ).

The central case around which the sensitivity analysis is performed uses the prior inaccuracy defined by the standard errors in table 3.2 ($\delta = 1$), a significant amount of experimental bias ($\lambda_1 = \lambda_2 = 1$) and a moderate degree of structural similarity ($\rho = .5$).

Table 3.3 is designed to demonstrate the sensitivity of the estimates to the choice of δ, the factor that multiplies the prior covariance matrix. A large value of δ indicates relatively little confidence in the prior estimates. The first column of numbers in table 3.3 has the prior estimates which apply if the prior is taken to be exact, $\delta = 0$. Note that the estimates are then the same for both developed and developing countries. The last column contains the separate OLS regressions, applicable if the prior is fully diluted ($\delta = \infty$) but also if there is no structural similarity ($\rho = 0$) and no experimental contamination $\lambda = 0$.

The central case is reported in the column headed by $\delta = 1$. These numbers might be compared with the pooled OLS estimates reported in table 3.3. Although the Bayes estimates for the two subsets are very similar, they are very different from the pooled OLS estimates in the sense that eight of the fourteen Bayes estimates are outside the two-standard error confidence sets around the pooled OLS estimates. The Bayes estimate of the effect of government expenditure, in particular, is much smaller than the pooled OLS estimate.

Nothing very dramatic happens as the inaccuracy of the prior information is increased or diluted. In particular, diluting the prior does not uniformly increase the size of the coefficient on government expenditure. This table indicates generally that the greatest departures from the pooled OLS estimates occur when the prior estimates are held with great confidence (small δ). This need not have been the case, and is not the case for some of the coefficients.

Table 3.4 has the Bayes estimates for different values of the experimental bias parameters: λ_1 (developed) and λ_2 (developing). Note that the middle column of this table is the same as the middle column in table 3.3. For the reasons discussed above, a value of λ equal to one indicates that the prior and data will be equally weighted if the sample size is infinite. A value equal to 10 indicates that $10/11 = 0.91$ of the weight is placed on the prior. A value equal to .1 indicates that a weight of $0.1/1.1 = 0.09$ is placed on the prior.

The convergence effect is largest when λ_1 is large. Then the developed data are treated as noisy and consequently more weight is put on the relatively large estimate using the developing data. The estimates of the effect of government expenditures vary little with λ. One other pattern that emerges in table 3.4 is that the Bayes estimates generally depart more from the pooled OLS estimates as the values of λ increase. Again this seems to repeat the lesson from the previous table. The prior is substantially different from the pooled OLS estimate and changes that tend to put more weight on the prior have more dramatic effects on the estimates.

Table 3.5 indicates the effects of changing the degree of similarity of the developed and developing regressions. A value of $\rho = 0$ indicates complete dissimilarity and no pooling. A value of $\rho = .99$ indicates that the regression coefficients are virtually identical. The similarity of the estimates does increase with ρ. In the case of the convergence coefficient, it is the coefficient of the developing data set that does the most adjustment. For the government effect, both coefficients get larger with ρ.

This small sea of numbers challenges our cognitive ability. It would be very helpful if the data could focus attention on a subset. Marginal likelihoods offer one tool for focussing. These statistics point to parameters that are most favored by the data, in the same way that likelihood ratios would. The difference is that instead of maximizing the likelihood function, a marginal likelihood is found by integrating the likelihood function using the prior distribution as weights of integration.

Ratios of marginal likelihoods evaluated at different sets of parameter values are called Bayes factors. The Bayes factors comparing different values of the prior parameters with the central case are reported in tables 3.6, 3.7, and 3.8. These Bayes factors are all defined relative to the central model, which accordingly has the Bayes factor of one in each of the tables. The Bayes factors in table 3.6 point to a sharper prior distribution. With the grid of δ-values in the table, the maximum Bayes factor occurs at $\rho = .5$, suggesting that the prior standard errors could be reduced by fifty percent.

It is more difficult to improve upon the central case by varying either the experimental bias parameters. The Bayes factor in table 3.7 is maximized at 56.1 when there is no experimental contamination, $\lambda = 0$, in both data sets. In this display the Bayes factors are slightly higher above than below the diagonal, indicating a very slight preference for relatively accurate developing data set. This result presumably comes from the (relative) similarity of the prior and developing OLS estimates.

The data information about the degree of similarity is the weakest. The Bayes factors in table 3.8 grow slightly with the degree of similarity, attaining the level of 2.14 when $\rho = .99$.

Generally a data analysis must combine estimation and sensitivity analysis. When the data are adequately informative, estimation is the preferred approach. These Bayes factors select ρ, the prior distrust parameter, with an adequate degree of precision. But neither the experimental bias λ nor the structural similarity can be estimated with adequate accuracy and the sensitivity analysis that was previously discussed comes into play. By limiting the sensitivity analysis to parameter values with adequately high Bayes factors, we can in this case narrow down the estimates very precisely.

4 Conclusion

This article proposes and illustrates an econometric method for pooling two noisy data sets. The pooled estimates depend on parameters measuring doubt about the prior, experimental contamination and structural similarity. Implicit or explicit reference must be made to these three concepts regardless of the method that is used to pool data. The explicit mathematical reference to these concepts in this paper is intended to facilitate the discussion about the assumptions that underlie a pooling exercise. The method is illustrated by a study of the determinants of the growth rates of developed and developing countries. These two data subsets yield estimates that are very similar, which makes the results not very sensitive to the form of pooling and which points toward low experimental bias and a high degree of similarity.

Table 3.1:
Variable Names and Descriptive Statistics

NAME	DEFINITION	MEAN	STD.DEV.	MIN	MAX
GR6085	Growth rate of real per capita GDP 1960–1985	0.022	0.019	-0.017	0.074
GDP60	1960 value of real per capita GDP (1980 base year, $ 1,000)	1.917	1.813	0.208	7.380
SEC60	1960 secondary-school enrollment rate	0.226	0.214	0.010	0.860
PRIM60	1960 primary-school enrollment rate	0.785	0.312	0.050	1.440
GOVEXP	Average from 1970 to 1985 of the ratio of real government cons. (excl. of defense and education) to the real GDP	0.107	0.053	0.000	0.245
REVCOUP	Number of revolutions and coup per year (1960–1985 or sub-sample)	0.180	0.231	0.000	1.150
ASSASS	Number of assassinations per million population per year (1960–1985 or sub-sample)	0.226	0.462	0.000	2.850
PRI60DEV	Magnitude of the deviation 1960 PPP value for the investment deflator from the sample mean	0.001	0.340	-0.494	1.827

Table 3.2:
OSL and Prior Estimates

Variables	OLS Estimates			
	Prior	Pooled	Developed	Developing
Gdp60	-0.01	-0.0074	-0.005	-0.011
	(0.01)	(0.001)	(0.001)	(0.002)
Sec60	0.1	0.0327	0.025	0.023
	(0.1)	(0.011)	(0.010)	(0.020)
Prim60	0.05	0.0224	0.022	0.027
	(0.05)	(0.006)	(0.014)	(0.008)
Govexp	-0.01	-0.1332	-0.042	-0.161
	(0.01)	(0.029)	(0.040)	(0.034)
Revcoup	-0.1	-0.0236	-0.019	-0.025
	(0.1)	(0.007)	(0.041)	(0.008)
Assass	-0.01	-0.0028	0.000	-0.002
	(0.01)	(0.003)	(0.006)	(0.004)
Pppidev	0	-0.0067	0.004	-0.008
	(1)	(0.004)	(0.013)	(0.005)

Table 3.3:
Sensitivity Respect to Prior Accuracy
($\lambda = 1$, $\rho = 0.5$)

VARIABLES	PRIOR	d=0.1	d=0.5	d=1	d=2	d=10	OSL
DEVELOPED							
GDP60	-0.01	-0.0092	-0.0077	-0.0076	-0.0075	-0.0075	-0.005
	(0.01)						(0.001)
SEC60	0.1	0.0693	0.0561	0.0553	0.0548	0.0550	0.025
	(0.1)						(0.010)
PRIM60	0.05	0.0458	0.0432	0.0427	0.0432	0.0428	0.022
	(0.05)						(0.014)
GOVEXP	-0.01	-0.0100	-0.0108	-0.0127	-0.0207	-0.0162	-0.042
	(0.01)						(0.040)
REVCOUP	-0.1	-0.0768	-0.0555	-0.0530	-0.0505	-0.0517	-0.019
	(0.1)						(0.041)
ASSASS	-0.01	-0.0093	-0.0057	-0.0046	-0.0042	-0.0042	0.000
	(0.01)						0.006
PPPIDEV	0	-0.0014	-0.0008	-0.0008	-0.0007	-0.0008	0.004
	(1)						(0.013)
DEVELOPING							
GDP60	-0.01	-0.0096	-0.0086	-0.0085	-0.0086	-0.0085	-0.011
	(0.01)						(0.002)
SEC60	0.1	0.0741	0.0527	0.0501	0.0477	0.0487	0.023
	(0.1)						(0.020)
PRIM60	0.05	0.0457	0.0456	0.0462	0.0481	0.0470	0.027
	(0.05)						(0.008)
GOVEXP	-0.01	-0.0100	-0.0113	-0.0142	-0.0266	-0.0196	-0.161
	(0.01)						(0.034)
REVCOUP	-0.1	-0.0608	-0.0497	-0.0492	-0.0478	-0.0486	-0.025
	(0.1)						(0.008)
ASSASS	-0.01	-0.0091	-0.0057	-0.0048	-0.0047	-0.0046	-0.002
	(0.01)						(0.004)
PPPIDEV	0	-0.0020	-0.0033	-0.0033	-0.0033	-0.0033	0.008
	(1)						(0.005)

Table 3.4:
Sensitivity with Respect to Experimental Bias
($\rho = 0.5$, $\delta = 1$)

	$\lambda_1 = 0.1$			$\lambda_1 = 1$		
VARIABLES	$\lambda_2 = 0.1$	$\lambda_2 = 1$	$\lambda_2 = 10$	$\lambda_2 = 0.1$	$\lambda_2 = 1$	$\lambda_2 = 10$
DEVELOPED						
GDP60	-0.0058	-0.0058	-0.0058	-0.0075	-0.0076	-0.0076
SEC60	0.0326	0.0341	0.0353	0.0473	0.0553	0.0628
PRIM60	0.0374	0.0373	0.0373	0.0424	0.0427	0.0428
GOVEXP	-0.0129	-0.0127	-0.0118	-0.0129	-0.0127	-0.0118
REVCOUP	-0.0286	-0.0312	-0.0335	-0.0442	-0.0530	-0.0609
ASSASS	-0.0019	-0.0022	-0.0024	-0.0039	-0.0046	-0.0053
PPPIDEV	0.0001	0.0001	0.0001	-0.0017	0.0008	0.0001
DEVELOPING						
GDP60	-0.0081	-0.0080	-0.0079	-0.0082	-0.0085	-0.0088
SEC60	0.0209	0.0439	0.0636	0.0221	0.0501	0.0764
PRIM60	0.0450	0.0446	0.0440	0.0453	0.0462	0.0465
GOVEXP	-0.0146	-0.0142	-0.0123	-0.0146	-0.0142	-0.0123
REVCOUP	-0.0188	-0.0430	-0.0629	-0.0197	-0.0492	-0.0757
ASSASS	-0.0023	-0.0041	-0.0059	-0.0026	-0.0048	-0.0072
PPPIDEV	-0.0063	-0.0030	0.0004	-0.0064	-0.0033	-0.0005

	$\lambda_1 = 10$		
VARIABLES	$\lambda_2 = 0.1$	$\lambda_2 = 1$	$\lambda_2 = 10$
DEVELOPED			
GDP60	-0.0089	-0.0092	-0.0095
SEC60	0.0593	0.0744	0.0900
PRIM60	0.0469	0.0479	0.0486
GOVEXP	-0.0126	-0.0124	-0.0115
REVCOUP	-0.0575	-0.0736	-0.0894
ASSASS	-0.0060	-0.0073	-0.0088
PPPIDEV	-0.0030	-0.0016	-0.0003
DEVELOPING			
GDP60	-0.0083	-0.0090	-0.0096
SEC60	0.0232	0.0557	0.0891
PRIM60	0.0456	0.0476	0.0492
GOVEXP	-0.0145	-0.0140	-0.0121
REVCOUP	-0.0204	-0.0550	-0.0889
ASSASS	-0.0028	-0.0057	-0.0088
PPPIDEV	-0.0065	-0.0035	-0.0006

Table 3.5:
Sensitivity Respect to Regression Similarities
($\lambda = 1$, $\delta = 1$)

VARIABLES	$\rho = 0$	$\rho = 0.5$	$\rho = 0.9$	$\rho = 0.99$
DEVELOPED				
GDP60	-0.0076	-0.0076	-0.0077	-0.0078
SEC60	0.0645	0.0553	0.0490	0.0475
PRIM60	0.0427	0.0427	0.0435	0.0438
GOVEXP	-0.0109	-0.0127	-0.0141	-0.0144
REVCOUP	-0.0627	-0.0530	-0.0467	-0.0454
ASSASS	-0.0055	-0.0046	-0.0042	-0.0041
PPPIDEV	-0.0001	-0.0008	-0.0019	-0.0022
DEVELOPING				
GDP60	-0.0091	-0.0083	-0.0080	-0.0078
SEC60	0.0568	0.0501	0.0475	0.0474
PRIM60	0.0479	0.0462	0.0444	0.0439
GOVEXP	-0.0138	-0.0142	-0.0144	-0.0144
REVCOUP	-0.0563	-0.0492	-0.0458	-0.0453
ASSASS	-0.0059	-0.0048	-0.0042	-0.0041
PPPIDEV	-0.0036	-0.0033	-0.0026	-0.0023

Table 3.6:
Bayes Factors for Different Values of Prior Accuracy

Delta	Bayes Factor
0.10	3.99E-10
0.20	2.35E+00
0.30	2.00E+02
0.40	3.59E+02
0.50	2.00E+02
1.00	1.00E+00
2.00	3.95E-04
10.00	1.94E-13

Table 3.7:
Bayes Factors for Different Values of Experimental Bias

$\lambda 1$ / $\lambda 2$	0	0.1	1	10
0	56.1084	41.5369	6.3868	0.0295
0.1	41.2674	30.9494	5.0295	0.0244
1	6.0710	4.8100	1.0000	0.0062
10	0.0241	0.0201	0.0053	4.32E-05

Table 3.8:
Bayes Factors for Different Values of Regression Similarity

ρ	Bayes Factor
0.00	0.66
0.50	1.00
0.90	1.80
0.99	2.14

References

Aigner, Dennis J., Hsiao, Cheng, Kapteyn, Arie and Tom Wansbeek (1984): "Latent Variables in Econometrics." In: Handbook of Econometrics, Volume II, eds. Zvi Griliches and Michael D. Intrilligator, Amsterdam: North-Holland Publishing Company, pp. 1321–1393.

Aigner, Dennis and Edward E. Leamer (1984): "Estimation of Time-of-Use Pricing in the Absence of Experimental Data: An Application of the Methodology of Data Transfer." *Journal of Econometrics*, **26**: 205–227.

Barro, Robert J. (1991): "Economic Growth in a Cross Section of Countries." *Quarterly Journal of Economics*, 407–443.

DeMouchel, W.H. and J.E. Harris (1981): "Bayes and Empirical Bayes Methods for Combining Cancer Experiments in Man and Other Species." Technical Report No. 24, Department of Mathematics, M.I.T., February.

Klepper, Steven and Edward E. Leamer (1984): "Consistent Sets of Estimates for Regressions with Errors in All Variables." *Econometrica*, **52**: 163–183.

Leamer, Edward E. (1974): "False Models and Post-Data Model Construction." *Journal of the American Statistical Association*, **69**: 122–131.

Leamer, Edward E. (1978): Specification Searches, New York: John Wiley and Sons, Inc.

Leamer, Edward E. (1991): "Measurement Errors and the Convergence Hypothesis." Anderson Graduate School of Management.

Patefield, W.M. (1981): "Multivariate Linear Relationships: Maximum Likelihood Estimation and Regression Bounds." *Journal of the Royal Statistical Society, Series B*, **43**: 342–352.

Chapter 4

An Econometric Model for Prices and Wages with Respect to the Economic Reform in Czechoslovakia

JAN VEVERKA

1 Introduction

The initial stage of the construction and application of price models in Czechoslovakia dates back to the years 1960 to 1965. In the course of time they were improved and represented for the central government an important instrument for the preparation and control of administrative intervention in prices. Leaving aside the purpose for which these models were used, they stimulated the creation of a comparatively solid data base, the compilation of programs for computer processing, and especially initiated a systematic study of problems connected with price and wage modeling.

The preparation of the economic reform required answers to questions, never formulated by central planning. We have to admit, that at the beginning we tried to find the answers by applying the former economic models. Naturally, we could not be successful. Models for an administrative price control had to be highly disaggregated in order to supply information for most addressed control – but once determined, prices and wages remained constant for quite a long period. That is why equations transforming the consequences of the price development into wages with an additional time lag could not be of any significance to them. Under the conditions of a price liberalization, the mathematical formalization of such relations, on the contrary, belongs to the decisive premise to any reasonable prediction. We were aware of the fact that a prediction of the price and wage development under the new conditions could be done only through some adaptations of an econometric model compiled, for a market economy that at least with respect some features was similar to Czechoslovakia.

2 The Economic Model

The literature describing compiled, tested and applied models is fairly manifold. After a detailed study, the optimal choice of the "foundation walls" for the Czechoslovak wage and price model seemed to indicate towards the model applied by the Bank of Finland called the BOF3 – Quarterly Model of the Finnish Economy. It was published by its authors Juha Tarkka and Alpo Willman in Helsinky, 1985. This model naturally provides a broader coverage of topics than only wages and prices. But the price and wage block stands out as a relatively independent part and permits us to separate it, if the rest of the variables are defined as exogenous quantities.

The price – wage block of the BOF3 is based on the division of the economy into open sectors and sectors sheltered from the foreign competition. The open sectors of the economy consist of forestry and manufacturing while the sheltered sectors comprise agriculture and services. In the Czechoslovak model adaptation forestry was left out. The derivation of the price and wage equations is based on the main assumptions as follows:

- According to the theory of a small open economy, the price elasticity of demand for products of the open sector is infinite in the long run. However, in the short term a firm has monopoly power, so that the short-run price elasticity of demand is finite.
- In the sheltered sectors the price elasticities of demand are finite both in short and long run.
- In both, the open and sheltered sectors (excl. agriculture) pricing is based on maximizing behavior. In agriculture the transmission of production costs to prices of final goods takes place in an administrative process.
- In all sectors, the production technology is assumed to be of the Cobb-Douglas type for value added components of output, but the gross production function in which labor and capital are interpreted as a joint factor of production, is of the Leontief type. Returns to scale are assumed to be constant and the technical process is assumed to be linked entirely to the value added component of production.

- Capital is assumed to be a fixed factor in the short run.
- In the formation of wages, the open sector (manufacturing) functions as a "wage leader", i.e. in the long run the rate of change in wages in sheltered sectors follows the rate of change in the open sector.
- Wages in the open sector react to the difference between the marginal productivity of labor and the marginal costs as well as the excess supply of labor.

The change in wages is divided into two components: changes in negotiated wages and wage drift. This division is significant if factors, on which changes in wage components are based, are not similar in all respects. This division also allows the use of negotiated wages as an autonomous decision variable. This seems reasonable, especially in short-run forecasting and simulation exercises. The analysis is based on the hypothesis that a rise in negotiated wages reduces the pressure on the wage drift.

These relations, of course, have to be confronted with the Czechoslovak reality. In our case, for instance, we cannot assert without reservation that there exists an open sector, which would have to adapt, although with a certain time lag, to prices on the world market. In our monopolistic conditions it is not even true that a change in the exchange rate has no direct influence on the prices of the domestic competitive production. At this point the question arises: which factors of the market are actually valid under our conditions? We have to admit that so far really very few.

The following three factors have become decisive for the price development under Czechoslovak conditions:

- the transition to world prices and convertible currencies in the trade with the Soviet Union and other member countries of the former CMEA (Council for Mutual Economic Assistance)
- the devaluation of the domestic currency and
- the price liberalization.

Our imports from the Soviet Union consist mainly of crude oil, gas and raw-materials for metallurgical production. The change of their prices did not affect the whole world market. It concerned only us, the former socialist countries. That is the reason why we left the bond development of world prices out of the Finnish model and replaced it by a costs equation using full coefficients taken from the INPUT-OUTPUT tables for Czechoslovakia.

The devaluation of the domestic currency was performed in several steps. The original reproduction costs of export and the official exchange rate for trade payments oscillated around 15 CSK/US$. The first devaluation increased it to 17 CSK/US$, and the second one to 24 CSK/US$. Finally the third one fixed it at 28 CSK/US$, equal to the tourist exchange rate, which is to fulfill the function of an "anchor" value.

For the present state of the Czechoslovak market we are fully entitled to say that a change in the exchange rate influences it in the same way as a price change of the material inputs. An integral part of the costs equation in our model is, therefore, an expression including the change in the exchange rate as well as the full coefficient of the import demand of our economy. There is, however, a considerable growth of domestic prices which caused a drop in the real value of the exchange rate to about 16 CSK/$ already during the first three months of 1991. That is why the question arises, whether we can regard the exchange rate as a purely constant and exogenous quantity or as a parameter that can be changed in time or even as an endogenous variable.

The most difficult problem in the price development forecast is the formalization and then quantification of the process taking place after the price liberalization. No foreign econometric model had to pay attention to such a phenomena and it was, therefore, not possible to just take over any suitable method. Our price model is based on the hypothesis that immediately after the liberalization of the prices, manufacturers will try to gain a profit which will cover their short-term needs, but also secure means for their future development. Some foreign studies indicate that for the absorption of new technologies offered on the world market the enterprises need resources amounting at 12 to 14 % of the value of their production. Under conditions of an extremely restrictive policy, enterprises will attempt to accumulate such resources by means of their own profits. The original profit value after taxation, that enterprises had at their disposal before lifting the price determination, should – with a view to the change in the profit tax – therefore, rise to two to three times its former value. The profit reduction caused by lower demand and its development over time must result from an independent study. It is obvious that it is just this quantity which reflects a multitude of opinions on the success of the reform process in our country. What was mentioned in connection with the development of profits, applies also to the estimation of the development of the production's material demands which – compared to the developed countries – is a 25 % to a 30 % higher. It also holds for labor productivity which, in the same comparison, does not even reach quite half the value.

The equation system adopted from the BOF3 actually fulfills the basic function expected from the model as such. These equations transfer an exogenous price impulse into the individual sectors, transform it into consumer prices and expose the population's behavior regarding demands for wage increases. This part of the wage development is connected with another one resulting from changes in labor productivity. The quantification process is returned to price correction till the equilibrium of the entire modelled system is attained.

We cannot imagine that the form and the parameters of such a system of equations could be derived from our own domestic experience. The past stopped to be valid and we are glad that it did. We hope, that the stage at the present, will not last too long. And we believe that our future could be similar to the situation in which the standard economies of the European countries are now. It can be expected that people react fairly similarly to price developments wherever they may take place and that they start to voice their demands with more or less the same delay. Perhaps even the lags of the value changes in the model quantities will not differ very much. The coefficients of the equations in model BOF3 were derived from time series nearly twenty years long. It can, therefore, be presumed that they reflect real relations and not phenomena appearing by chance.

The equation system of the BOF3 model had also to be modified with respect to the social agreement on the development of real wages. According to this system, enterprises will get every three months information on the permitted wage growth without having to face sanctions taken against them. The decisions about the wage increase is taken by the government on the basis of past developments and its deeper sense lies in insuring that the drop in real wages should not be higher than the agreed upon limit.

3 Actual Forecasting Performance

The model works with a time variable of three months. For a situation like the Czechoslovak one, this is an ideal solution. For the entire prediction period the employer's social security contributions rate, the interest rate and corporate tax rate are assumed to be constants. The main exogenous quantities for each time interval are the exchange rate, the price indices of crude oil, raw materials and gas, the GDP growth rate, unemployment rate and the labor productivity rate.

The first quantifications have been performed with the help of this model in autumn 1990. Their aim was to estimate the potential pressures on prices and wages starting on January 1st 1991, the date of the beginning of price liberalization. But this pressures partially influenced prices during 1990 to. Therefore we presume that December 1989 = 100 %. The results of the model calculation shown in figure 4.1 are actually a pessimistic scenario as they do not presume the start of an adaptation process. Under these presumptions the price level would rise after two years for manufacturing by about 180 to 190 %, for agriculture by about 215 to 225 % and for services by 230 to 240 %. Real wages would drop by 30 to 35 %. Simultaneously with this task we also performed a number of simulations with the aim to discover the risks of a rapid inflation change. It was clearly shown that the future development will depend on the extent to which we shall be able to eliminate the monopolistic behavior of enterprises, i.e. on the degree to which their profits will be kept at a reasonable level. The marginal border-line for the convergence or divergence of the price development – with a corporate tax rate of 55 % – is an approximately 14 % net profit rate. The period of one year, however, does not allow us to identify the real development from a long-term point of view.

Towards the end of 1990 some facts accepted for the first estimate changed (among others, an agreement on the maximum drop in real wages was reached) and a sort of consistent (or even optimal) idea on the nature and course of the adaptation process has been formed. Both estimates with a pessimistic as well as an optimistic correction (including the calculation of the average) for manufacturing are shown in figures 4.2, 4.3 and 4.4. After three years the price development will most probably stabilize with 130 to 140 % rise and at that time real wages will still be 5 to 10 % lower than their initial level at the end of 1990 after having grown by 110 to 120 % in the course of this development.

So far this prediction can only be compared with one single observation actually corresponding to the very beginning of the diagram. For the first three months, the statistical office published a price rise in manufacturing amounting to 73,9 % while the estimation covered an interval of 68 to 75 % and a price rise of 23,9 % for services compared to an estimation of 20 to 30 %. The estimation for the real drop of real wages lies between 30 and 35 % in accordance with the model forecast. As to the individual factors participating in the total price growth for manufacturing, about 20 % was caused by an increase in prices of raw-materials from the Soviet Union. Another about 20 % is the result of the devaluation of the domestic currency and the last 20 to 25 % was caused by price liberalization. Material inputs (prices and devaluation) and liberalization participate to about 50 % in the price rises of the services.

4 Informal Problems with Forecasting

The estimation of the price, wages and general development of our economy is connected with an extremely broad spectrum of problems. It seems that those included in econometric frameworks belong to the group of simpler and easier identifiable ones. The phenomena we are facing in the course of price liberalization and which we shall face, are, as can be observed, not only a reflection of the newly appeared situations, but also have their deep roots in the past. From the deformed so-called consumer-supplier relations and the relations between particular subjects, individuals transfer a feeling of injury, a feeling that they suffered and injustice was committed on them, because somebody else was given the chance of gaining unjustified advantage. That is why we can express the opinion that the price and wage development will not only represent the result of real changes in the development of costs, but will reflect the acquired and newly shaping traits of the participants of the few just established market. They will probably become more of an ethical rather than economical category.

The only "competitive" anti-pole, influencing their determination, will probably very often represent mere subjective extent to which a fraud can be committed against the buyer side. Many examples of the behavior of the still existing monopolies in production, trade and certain groups of private entrepreneurs are convincing us that we have moved more in the direction of a "Balcanized" market type than towards relations between buyers and sellers which we admire during our visits to West-European countries.

Resulting from the lack of many legal measures, an unmerited enrichment of a certain group of the population is spreading without the entire society's participation in these resources by ways of taxes and duties. Other groups – with no fault of their own – are approaching the critical threshold of the life minimum.

With the entry of certain groups, rather adventurously enterprising than entrepreneurial, taking place between production and consumer, the impression appears that the price liberalization is starting to introduce a competitive atmosphere and the citizen is forced to accept the idea that the reason for our high prices lies in the trade network. In reality, practically no competition has so far entered the production side and prices are high mainly due to the fact, that we are producing with high costs and low labour productivity.

The level of costs at the beginning of our reform, the costs with which we are "opening" our country to the world, is characterized by the values given in the following table. It takes the form of the first quadrant of a classical input-output balance disintegrated into the 1st, 2nd and 3rd sector. These various values express how many times our material consumption is higher than that in developed industrial countries.

Table 4.1:
Sector Requiring Supplies From the Input-Output Table

		I	II	III
Sector	I	1.3	1.9	1.2
	II	0.8	1.5	1.0
supplying	III	0.9	0.4	0.9

Compared to these countries, the internal material consumption of our primary sector is 30 % and that of the secondary one 50 % higher. The secondary sector requires 90 % more products from the primary sector. On the other hand, we are lagging behind in goods supplied by the tertiary sector. That is mainly true for the secondary sector which receives only 40 % of its supplies from the tertiary sector. But the service sectors itself, also requires much more resources produced by the primary sector.

A sufficient idea about the size of deformations in present retail commodity prices can be given by a simple nominal comparison with commodity prices on the German market, represented with the prices of services paid by an average family in the German as part of its expenditures. In the main group of consumer's expenditures these nominal relations were in 1987 approximately of the following nature:

Food	3.6
Clothing and shoes	5.3
Industrial goods	7.6
Rent	1.0
Heating and lighting	1.4
Services	1.3

The reasons for these differences result from the irrational subsidies and tax policy of the state. The extent to which we shall be able to master the social consequences of their elimination and to which we shall succeed in creating an environment in which production will have no choice but to reduce its costs, will decisively influence the reform process, including the price and wage development.

5 Conclusions

An ample elaboration of the BOF3 tempts us to adapt more of its parts to the Czechoslovak conditions. To make it quite clear – adapt to conditions that are connected with the transformation of this country to a market economy. In this process we have to achieve fundamental changes in the labor productivity, in the energy and material requirements of production – and all this in connection with a change in the production structure. This is obviously the reason why we will not succeed to adapt the model in blocks like employment and labor force, inventories, output, potential output, imports, exports, consumption and investment. More hopeful might seem to form blocks of financial nature, such as taxes and public finances, and balance of payments. Maybe that even this intention is too daring. Such efforts might benefit only when the economy is stabilized and shows at least basic features of a market system.

The inspiring effect of the publication "The BOF3 Quarterly Model of the Finnish Economy" is, however, in any case extremely strong. It helped me to overcome obstacles which I would have mastered only with great difficulties.

Reference

Tarkka, J. and A. Willman (1985): "The BOF3 Quarterly Model of the Finnish Economy." Bank of Finland, Helsinky.

Figure 4.1:
Forecast of Pressure

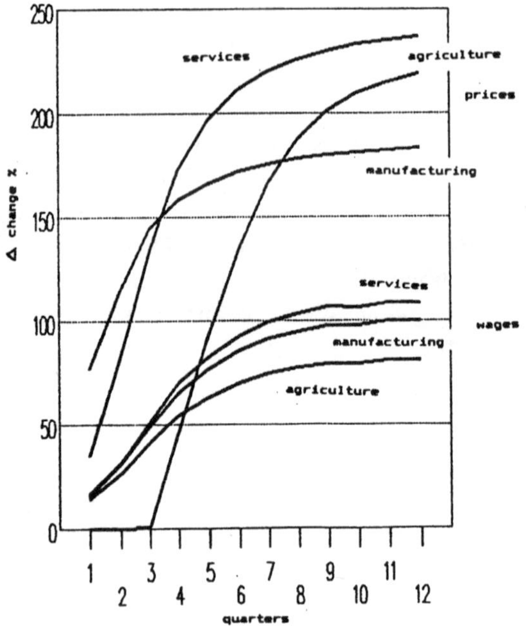

Figure 4.2:
Manufacturing – Prices

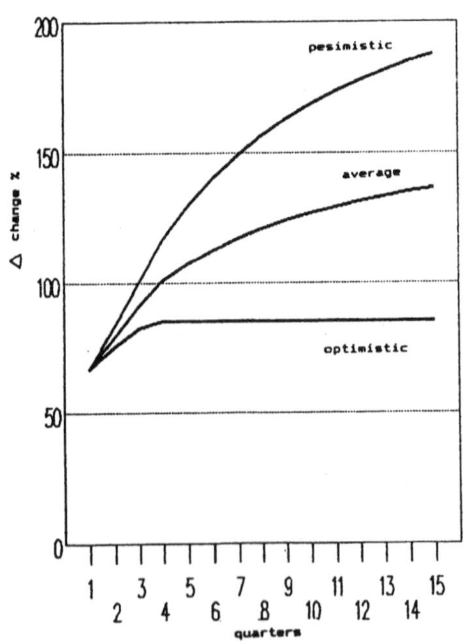

Figure 4.3:
Manufacturing – Wages

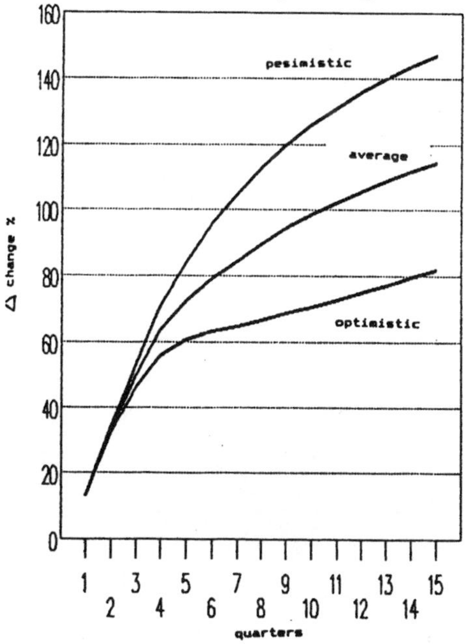

Figure 4.4:
Manufacturing – Real Wages

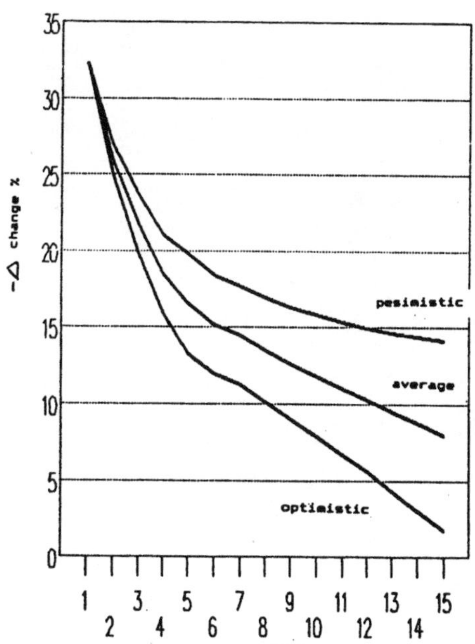

Chapter 5

Using Extraneous Information to Estimate Time Series Models. A Review of Approaches Applied in Market Response Modeling

HARALD HRUSCHKA

1 Introduction

A response model reproduces the dependence of one variable on a set of different variables. The independent variables of market response models consist of marketing instruments (like prices, advertising budgets, sales force size, distribution) and environmental variables (like GDP, weather), their dependent variables are marketing goals (like market volume, sales, market share). Market response models represent essential components of marketing decision models that serve to either assess normative consequences of given marketing policies or to determine optimal marketing policies (i.e. levels of marketing instruments) (Parsons and Schultz [1978], Lilien and Kotler [1983], Hanssens, Parsons, and Schultz [1990], Hruschka [1995]).

The standard procedure is to estimate parameters of market response models by means of standard econometric and times series analysis techniques. As a rule, historical data for a common span of time referring to a fixed set of observational units (e.g. products or brands) are used. Other information, acquired for example from decision makers or (in some regard) comparable studies, is typically ignored.

Of course, such an exclusive reliance on a single set of historical data is not without problems (Naert and Weverbergh [1981]). Part of the data required may not be available. A simple reason could be that they had not been recorded. It is also possible that some of the marketing instruments were not used in the past (e.g. a company adding some form of computer-based selling to its marketing-mix or introducing a new product).

Often variability of data is too low for standard econometric and time series techniques, as companies are not willing to execute more important changes of levels of marketing instruments fearing lower profits. In such cases unreliable parameter estimates result. This problem is obviously aggravated the shorter the available time series are.

These considerations suggest to resort to estimation techniques that obtain information both from a common set of historical data and from extraneous information. It can be hoped that some of the problems mentioned above might be overcome this way.

2 Sources of Extraneous Information

In the field of operations research/management science subjective estimates acquired from decision makers constitute an important and (as the literature on applications shows) frequently used type of information.

As is well known, subjective estimates may be prone to motivational (Spetzler and Stael v. Holstein [1975]) or cognitive biases (Tversky and Kahnemann [1975]). In the first case, the subject modifies his judgment in order to influence decisions that are based on it. Motivational biases result from erroneous integration of information by decision makers (e.g. being overly influenced by more recent events). Experimental evidence shows that errors of subjective parameter estimates are higher if the level of a marketing instrument lies outside the range familiar to the decision maker (Chakravarti, Mitchell, and Staelin [1979]).

An alternative to judgments of the decision maker is strictly empirical extraneous information gained by different studies that analyzed at least one sample different from the current one. Results of studies estimating elasticities of marketing instruments from large databases like those of the PIMS project (covering 3000 businesses) might be useful in this respect. Table 2 contains mean elasticities of marketing instruments with regard to sales obtained by Hagerty Carman, and Russel [1988].

Table 5.1:
Elasticities from PIMS Data

Instrument	Mean Elasticity	Standard Deviation
Quality	0.344	0.528
Price	-0.985	1.969
Advertising	0.003	0.105
Promotion	0.008	0.082
Sales Force	0.304	0.255

A problem with using results of past studies is their often doubtful similarity to the new modeling or forecasting tasks. Meta-analysis generalizes results of past studies (Farley and Lehmann [1986]). It compares similar but not necessarily identical estimates of quantities measured in different settings that are seen as treatments of a natural experiment. In most cases ANOVA (or some related method) is used to explain variation of results by interstudy or intermodel differences. Meta-analyses of central results of market response studies seem to be interesting sources of information to the modeler plagued by short or unreliable data series.

The subject of the meta-analysis of Tellis (Tellis [1988]) is aggregated price response. Average elasticity of price over 424 models from 42 studies amounts to -1.76. About 29 % of the variation in elasticity is explained by factors like data interval, stage in the product life cycle or product class.

A meta-analysis of Assmus et al. [1984] processes 128 models from 22 studies. They compute an average short-term advertising elasticity of 0.22. Short term elasticity lies in the interval $[0.0, 0.49]$. About 50 % of variation in short-term advertising elasticity between models is explained by the corresponding ANOVA model.

Sultan et al. [1990] meta-analyze the results of diffusion studies published in 15 articles comprising a total of 213 sets of parameters. Diffusion means the development of first purchases of a new product over time by a population. Especially in the early periods, forecasting of purchases of a new product is very difficult, as parameter estimates are extremely uncertain due to lack of data. Therefore, extraneous information on parameters is of great benefit. A rather general formulation of diffusion is (Mahajan and Peterson [1985]):

$$\frac{dN(t)}{dt} = g(t)[N_{max} - N(t)] \tag{5.1}$$

Here $dN(t)/dt$ is the rate of diffusion at time t, $N(t)$ is the cumulative number of adopters at time t, N_{max} the total number of potential adopters, $g(t)$ the adoption rate.

A popular functional form of $g(t)$ is (Bass [1969]):

$$g(t) = p + q[F(t)] \tag{5.2}$$

where p and q measure innovation and imitation effects, respectively and $F(t)$ is a generalized logistic curve.

According to this meta-analysis p averages 0.03, q 0.38 with considerable variance. Up to 50 % of the variance of these two parameters is explained by differences of the studies that concern region, product type (industrial vs. consumer durable).

Such meta-analyses provide expected values and variances of parameter of market response models. These values could serve as input to econometric methods combining extraneous and sample information. Such input should especially be fruitful in situations where data are short or unreliable.

3 Exact Extraneous Information

Informal approaches combine estimates based on different information sources in an ad-hoc manner. Most of these approaches acquire subjective estimates by interviewing decision-makers.

The approach developed by Lambin [1972] starts with estimating an econometric model on the basis of a single historical data set. The parameters computed this way are subsequently adjusted by multiplying them with subjectively estimated indices whose reference value is unity. In the cited study the parameter of advertising is multiplied by a so-called advertising creativity and communication value index that should reflect experience of an advertising manager on the influence of creativity aspects. Besides general problems of subjective estimation discussed above a further weakness is the exclusion of any relationship between the variance of the econometric estimate and the adjusted value.

So called decision calculus models (Little [1970]) are frequently used in marketing decision support systems. To obtain parameters, decision makers are asked to indicate the output value of market share or sales they expect for each of a number of different sets of input values for independent variables like advertising budget, price etc. To these subjective point estimates a prespecified deterministic function (said to reproduce the market response model of the decision maker) is fitted.

The use of point estimates clearly falsely eliminates some of the uncertainty in the response, but as a rule repeated responses to measure error are not demanded from a decision maker. Usually the number of data values is low relative to the number of parameters, which makes sensitivity analyses of parameter values impossible. Little (Little [1970]) even determines a four parameter function starting from four subjective values. Of course, any function with four parameters fits these data, even if it is meaningless and there is no way to decide between different functional forms. Proponents of decision calculus typically put most emphasis on this subjective estimation exercise. However, in some applications parameter values are adjusted on the basis of historical data (Little [1975]). This way, the decision calculus approach may finally come close to the usual econometric procedure.

Besides these ad-hoc procedures there are some econometric methods incorporating extraneous information that is considered to be exact (Judge, Griffiths, Hill, and Lee [1980]). The most basic approach enters (averages of) parameter values obtained subjectively or on the basis of other data sets into response functions (Claycamp and Liddy [1969]).

For linear models (or models that may be transformed into linear form) the parameter vector is partitioned into two subvectors:

$$\boldsymbol{q} = [\boldsymbol{X}_1 \boldsymbol{X}_2] \begin{bmatrix} \boldsymbol{\beta}_1 \\ \boldsymbol{\beta}_2 \end{bmatrix} + \boldsymbol{\epsilon} \tag{5.3}$$

Now the outside estimate $\boldsymbol{b}_1$ for $\boldsymbol{\beta}_1$ is substituted to give the following model:

$$\boldsymbol{q} - \boldsymbol{X}_1 \boldsymbol{b}_1 = \boldsymbol{X}_2 \boldsymbol{\beta}_2 + \boldsymbol{\epsilon} \tag{5.4}$$

This way estimation is alleviated as the number of parameters to be determined is reduced.

A slight extension of this approach, called restricted least squares consists in considering extraneous information on linear combinations of parameters. Examples for such kind of information are the sum of a subset of parameters or equality of some parameters. Such information can be expressed by:

$$\boldsymbol{R\beta} = \boldsymbol{r} \tag{5.5}$$

$\boldsymbol{R}$ is a given matrix of order $s \times k$ with $s \leq k$ and $\boldsymbol{r}$ a given vector with s elements. Application of Langrangean minimization to this problem yields the optimal estimator.

Inequality restricted least squares allows consideration of exact extraneous information in form of linear restrictions:

$$\boldsymbol{R\beta} \leq \boldsymbol{r} \tag{5.6}$$

This method can be applied, for instance, if a parameter value is known to lie in a certain interval (i.e. minimum and maximum values are given, respectively) or known to be positive (negative). Estimators are derived by applying the Kuhn-Tucker conditions for nonlinear inequality-restricted problems.

Of course, extraneous information in form of linear (in-)equalities can be incorporated in nonlinear models in an obvious manner and estimated by a suitable estimation method (using e.g. reparametrization, Langrange multipliers, penalty functions).

Advantages of deterministic approaches lie in their computational and data-gathering simplicity. Implementations of marketing decision models may benefit from direct and easy involvement of decision makers. But all informal approaches ignore the uncertainty of extraneous information. This must be considered to be their most important weakness.

4 Imprecise Extraneous Information

4.1 Equation-Based Approaches

If information on linear restrictions of parameters is no longer considered to be exact, but to be subject to stochastic disturbances in form of a normally distributed vector $\boldsymbol{\mu}$ equal to equation 5.5 is modified to:

$$\boldsymbol{R\beta} + \boldsymbol{\mu} = \boldsymbol{r} \tag{5.7}$$

The resulting model may be written as:

$$\begin{bmatrix} \boldsymbol{q} \\ \boldsymbol{r} \end{bmatrix} = \begin{bmatrix} \boldsymbol{X} \\ \boldsymbol{R} \end{bmatrix} \boldsymbol{\beta} + \begin{bmatrix} \boldsymbol{\epsilon} \\ \boldsymbol{\mu} \end{bmatrix} \tag{5.8}$$

Mean vector and covariance matrix of $[\boldsymbol{\epsilon}'\boldsymbol{\mu}']$ are:

$$\boldsymbol{E}\begin{bmatrix} \boldsymbol{\epsilon} \\ \boldsymbol{\mu} \end{bmatrix} = \begin{bmatrix} 0 \\ \boldsymbol{\delta} \end{bmatrix}; \quad \boldsymbol{E}\begin{bmatrix} \boldsymbol{\epsilon} \\ \boldsymbol{\mu} \end{bmatrix}\begin{bmatrix} \boldsymbol{\epsilon} \\ \boldsymbol{\mu} \end{bmatrix}' = \sigma^2 \begin{bmatrix} \boldsymbol{I}_T & 0 \\ 0 & \boldsymbol{\Omega} \end{bmatrix} \tag{5.9}$$

The appropriate generalized least squares estimation technique for this kind of problem using an unbiased estimate for σ^2 is called stochastic restricted least squares or mixed estimation (Judge, Griffiths, Hill, and Lee [1980]).

Rao and Yamada [1988] estimate sales of a new brand as a function of a brand's own and competitive marketing efforts and cumulative sales. The four parameters of this model are expressed as linear functions of perceived product attributes. Coefficients of these linear functions are estimated on the basis of time series as well as survey data on perceptions of similar brands. Inserting perceptions of the new brand in these linear functions provides linear stochastic restrictions. The application of stochastic restricted least squares allows forecasting sales in spite of lack of data.

The approach recommended by Vanhonacker formally considers results obtained from meta-analysis (Vanhonacker, Lehman, and Sultan [1989]). Three linear regression equations describe the basic model, the meta-analysis model and the parameter uncertainty, respectively:

$$\boldsymbol{q} = \boldsymbol{X\beta} + \boldsymbol{\epsilon} \tag{5.10}$$

$$\boldsymbol{\beta} = \boldsymbol{Z\gamma} + \boldsymbol{u} \tag{5.11}$$

$$\boldsymbol{\gamma} = \boldsymbol{E}(\boldsymbol{\gamma}) + \boldsymbol{\mu} \tag{5.12}$$

Sultan et al. [1990] use the results of their a meta-analysis of diffusion studies. The authors start with expected values of p and q for US durables when forecasting product growth for a time series of air room conditioners. Mixed estimation appeared to be more robust than nonlinear least squares estimation, especially at the beginning of the series. After seven years the extraneous information almost had no impact.

Kennedy generalizes stochastic restricted least squares (Kennedy [1991]). The elements of $\boldsymbol{R}$ may now be nonlinear functions of an unknown parameter vector $\boldsymbol{\theta}$ for which an extraneous unbiased estimate $\boldsymbol{d}$ with covariance matrix $\boldsymbol{V}(\boldsymbol{d})$ is given.

The well-known Bass-Model (Bass [1969]) for new product growth serves Kennedy to demonstrate an application of this technique.

$$\begin{aligned} S(t) &= [p + qN(t)/N_{max}][N_{max} - N(t)] \\ &= [pN_{max} + (q-p)N(t)] - (q/N_{max})N^2(t) \end{aligned} \tag{5.13}$$

$S(t)$ indicates initial purchases at time t.

For time series data the following formulation may be used:

$$S(t) = a + bN(t-1) + cN^2(t) \tag{5.14}$$

The equalities $a = pN_{max}, b = q - p, c = q/N_{max}$ imply the nonlinear relationship $a + N_{max}b + N_{max}^2 c = 0$.

Given information on expected values and variance of either a or N_{max}, from which the nonlinear restriction just introduced may be computed, the other parameters of the Bass Model may be estimated by the extension of stochastic restricted least squares to nonlinear restrictions.

Another stochastic approach to incorporate extraneous information, Bayesian regression, reflects uncertainty about parameters by prior and posterior density functions.

Under certain assumptions the likelihood function l for linear regression is multivariate-normal:

$$l(\boldsymbol{Q}|\boldsymbol{X},\boldsymbol{\beta},\boldsymbol{\sigma}) \propto \sigma^{-n} \exp^{-\frac{1}{2\sigma^2}[(n-k)s^2+(\boldsymbol{\beta}-\boldsymbol{b})'(\boldsymbol{X}'\boldsymbol{X})(\boldsymbol{\beta}-\boldsymbol{b})]} \tag{5.15}$$

$\boldsymbol{b}$ is the OLS estimate of the regression coefficients and s^2 of σ^2.

The conjugate prior density of the multivariate-normal likelihood when σ is unknown is the normal-gamma density:

$$p(\boldsymbol{\beta},\sigma) \propto \sigma^{-n_0} \exp^{-\frac{1}{2\sigma^2}[(n_0-1-k)\sigma_0^2+(\boldsymbol{\beta}-\boldsymbol{\beta}_o)' s_0^2 \boldsymbol{V}_o^{-1}(\boldsymbol{\beta}-\boldsymbol{\beta}_o)]} \tag{5.16}$$

$\boldsymbol{\beta}_o$ a priori mean values of $\boldsymbol{\beta}$

$\mathbf{V}_0$ a priori covariance matrix of $\boldsymbol{\beta}$

σ_0^2 a priori mean of σ^2

n_0 integer, $n_0 > k+1$

The so-called hyperparameters $\boldsymbol{\beta}_o, \boldsymbol{V}_o, s_0^2, n_0$ must be set to numerical values. The higher the perceived reliability of the elements of $\boldsymbol{\beta}_o$ is, the lower the (diagonal) elements of $\boldsymbol{V}_o$ are set. s_0^2 measures the variance of the priori model. n_0 increases (relative to k) with confidence in the priori estimation.

Combining the likelihood and the priori density according to Bayes' theorem gives the posteriori density for $\boldsymbol{\beta}$ and $\boldsymbol{\sigma}$ whose parameters reflect total (sample and extraneous) information. The amount of extraneous information needed is greater than that of stochastic restricted least squares. Besides intervals for regression coefficients it consists of their priori covariances, the priori residual variance and n_0. An application of Bayesian regression to estimating a multiplicative market response model for a short data set and subjective information on elasticities is given in (Nenning [1980]).

4.2 Rule-Based Approaches

Conventional market response models consist of equational relationships among (for the most part) continuous (dependent and independent) variables. To handle imprecision of equational relationships, econometric techniques based on probability theory are used.

A different representation of market mechanism is based on rules instead of equations. Rules are conditional statements consisting of (at least) two components (Shortliffe [1984]):

- a premise, which is a conjunction of one or more conditions
- a conclusion

A rule might look as follows:

If the price of brand 1 is low and advertising expenditures of brand 1 are medium then sales of brand 1 are medium.

Most decision makers in the business domain are not accustomed to express their knowledge in terms of equations and their parameters. A collection of rules reflecting verbal qualitative knowledge, on the other hand, might ease communication with decision makers and especially formulation of their cognitive model about market response. Such an approach is similar in spirit to decision calculus modeling. Its aim is to reproduce the decision maker's cognitive model, too. But contrary to decision calculus models no fixed functional form is assumed. This greater flexibility should be more faithful to a decision maker's view of the market mechanism.

There is no reason to suggest that rules describing market response are deterministic (i.e. free from imprecision). The literature on expert systems and decision support systems contains quite a lot of different approaches to manage imprecision of rules (Ng and Abramson [1990]). The more familiar ones are based on Bayesian probability theory, Dempster-Shafer theory, fuzzy logic and certainty factors.

The present paper deals with a fuzzy logic approach. This approach is chosen because it has been applied to real marketing data (Hruschka [1988]). The procedure starts with a so-called linguistic model consisting of fuzzy variables and linguistic relationships between variables. The linguistic relationships are a collection of rules supplied by a decision maker.

The fuzzy variables are defined in terms of fuzzy sets. They correspond to different levels of the original variables (e.g. price is low, advertising expenditures are high etc.). The original variables correspond to dependent and independent variables familiar from market response modeling.

A fuzzy set A is equivalent to an ordered pair (Zadeh [1965]):

$$A = (x, u_A(x)) \quad , \quad x \in X \tag{5.17}$$

with $0 \leq u_A(x) \leq 1$, where X is a non-fuzzy (i.e. classical) support set, A is a fuzzy subset of X, $u_A(x)$ is the membership value of x in A.

Allowing gradual transition in membership distinguishes fuzzy set theory from classical (hard) set theory that only knows the two cases of complete membership and complete non-membership. Fuzzy membership functions may be obtained from frequency distributions of empirical data or interviewing decision makers indicating, for example, to what extent they regard certain values of sales as high.

Inference strategies process given rule sets (N is the number of rules). The most widespread inference strategy is modus ponens. It states that if the premises of a rule are true, it is valid to deduce that its conclusion is true.

Facts: A is $X(i)$

Rules: if A is $X(i)$ then B is $U(i)$, $i = 1, ...N$

Conclusion: B is $U(i)$

If $X(i)$ and $U(i)$ are fuzzy sets, such rules may be presented in form of fuzzy relations $R(i)$ (Pedrycz [1985]):

$$R = \cup_{i=1}^{N} R(i) \tag{5.18}$$

with $R(i) = X(i) \times U(i)$, i.e. the cartesian product of the fuzzy sets $X(i)$ and $U(i)$. Each cell of the multidimensional matrix R may be interpreted as implication with a fuzzy truth value (Pedrycz [1983]):

$$Y = \vee X(1) \wedge X(2) \wedge X(3) \cdots \wedge X(N) \wedge R \tag{5.19}$$

where Y is the linguistic output variable, $X(1), \cdots, X(N)$ linguistic input variables and R fuzzy relations.

High truth values indicate a high degree of belonging to the respective value of a linguistic output variable, in other words that the rule predicts this value well.

The application discussed here uses a widespread type of fuzzy inference, max-min. For each rule of a rule set the minimum of the membership values of its linguistic input variables and its fuzzy truth value is computed. This value indicates how well the rule matches the value of the linguistic output variable given the (fuzzy) input data. Such minima are determined for all rules referring to one value of a linguistic output variable (e.g. sales medium). The maximum of these minima serves as membership value for the value of the linguistic output variable, i.e. the best-matching rules determines the final result (in AI parlance, it fires). These computations are repeated for the other values of the output variable(s).

Hitherto I only explained the working of a market response model represented by a fuzzy rule set given values for the input (independent) variables. This corresponds to computing values of independent variables by solving this equations. In the application mentioned above extraneous information is a very simple linguistic market response model (5 rules) obtained from decision makers. Combination with a small data set of time series requires transformation of metric time series into values of linguistic variables on the basis of membership functions acquired from decision makers. Now a model fit between these values of the linguistic output variable (sales) and values computed by max-min composition of fuzzy rules and values of the linguistic input variables (actually prices of three brands of a product group and average monthly temperature) can be computed. Maximizing model fit for these observed linguistic variables is equivalent to a nonlinear optimization problem. Solving this optimization problem together with a heuristic search for new rules on the basis of partial derivatives may lead to improved market response models made up of fuzzy rules. In the application, simplification (aggregation) of the now 46 rules is necessary to alleviate interpretation.

The optimization and search procedure modifies rules if their truth values are low, i.e. a high truth value makes a rule resistant to change. A high truth value should therefore be chosen only if confidence in a rule is very high.

5 Conclusions

The greatest advantage of assuming exact extraneous information seems to be the ease of using appropriate methods. This article presented three approaches processing exact information, namely subjective indices, decision calculus models and appropriate econometric techniques. Their common weakness lies in the fact that they ignore the frequent imprecision of extraneous information.

One group of methods considering imprecise extraneous information treated here is equation-based. These methods are part of the econometrician's toolkit. In order of increasing information needs, stochastic restricted least squares and Bayesian regression are discussed. The other group of methods is rule-based.

Models of market response are no longer represented by equations, but by rules. Rules may be appropriate if prior knowledge must be rated as qualitative and rather imprecise. The fuzzy-logic approach introduced here does not require prior information on parameter values. Its rule set may be updated on the basis of incoming empirical information.

All applications sketched here concern market response modeling, which is a central theme of marketing research. Because of the typical lack of suitable data, consideration of extraneous information is a necessity, especially if medium and long range marketing decisions are to be prepared and controlled. Therefore the development of appropriate methods to support decision makers seems more than justified.

Hopefully, some of the topics and solutions discussed (forecasting new-product-growth, gaining extraneous information from meta-analysis of past empirical studies as well as rule-based approaches) may be of interest to neighboring fields that have to deal with similar problems caused by short and unreliable time series.

References

Assmus, G., J. U. Farley, and D. R. Lehmann (1984): "How Advertising Affects Sales. Meta-Analysis of Econometric Results." In: *Journal of Marketing Research*, **21**: 65–74.

Chakravarti, D., A. Mitchell, and R. Staelin (1979): "Judgment Based Marketing Decision Models. An Experimental Investigation of the Decision Calculus Approach." In: *Management Science*, **25**: 251–263.

Bass, F. (1969): "A New Product Growth Model for Consumer Durables." In: *Management Science*, **15**: 215–227.

Claycamp, H. J. and L. E. Liddy (1969): "Prediction of New-Product Performance. An Analytic Approach." In: *Journal of Marketing Research*, 414–420.

Farley, J. U. and D. R. Lehmann (1986): "Meta-Analysis in Marketing. Generalization of Response Models." Lexington Books, Lexington, MA.

Hagerty, M. R., J. M. Carman, and G. J. Russel (1988): "Estimating Elasticities with PIMS Data. Methodological Issues and Substantive Implications." In: *Journal of Marketing Research*, **25**: 1–9.

Hanssens, D. M., L. J. Parsons, R. L. and Schultz (1990): "Market Response Models." Econometric and Time Series Analysis, Kluwer Academic Publishers, Boston, MA.

Hruschka, H. (1988): "Use of Fuzzy Relations in Rule-Based Decision Support Systems for Business Planning Problems." In: *European Journal of Operational Research*, **34**: 326–335.

Hruschka, H. (1995): "Marktwirkungsfunktionen." To appear in: R. Köhler, B. Tietz and J. Zentes (Hrsg.): Handwörterbuch des Marketing. Poeschel, Stuttgart.

Judge, G. G., W. E. Griffiths, R. C. Hill, and T.-C. Lee, (1980): "The Theory and Practice of Econometrics." Wiley, New York.

Kennedy, P. (1991): "An Extension of Mixed Estimation with an Application to Forecasting New Product Growth." In: *Empirical Economics*, **16**: 401–415.

Lambin, J. J. (1972): "A Computer On-Line Marketing Mix Model." In: *Journal of Marketing Research*, 119–126.

Lilien, G. L. and Kotler, P. (1983): "Marketing Decision Making." Harper and Row, Cambridge, MA.

Little, J. D. C. (1970): "Models and Managers. The Concept of Decision Calculus." In: *Management Science B*, **16**: 466–485.

Little, J. D. C. (1975): "BRANDAID. A Marketing Mix Model." Part II. Implementation, Calibration and Case Study. In: *Operations Research*, **23**: 656–673.

Mahajan, V. and R. A. Peterson (1985): "Innovation Diffusion Models and Applications." Sage Press, Beverly Hills, CA.

Naert, P. A. and M. Weverbergh (1981): "Subjective versus Empirical Decision Models." In: R. L. Schultz and A. A. Zoltners (eds.): Marketing Decision Models, North Holland, New York, pp. 99–123.

Nenning, M. (1980): "Bayes'sche Inferenz bei der statistischen Auswertung von Marktdaten." In: *Zeitschrift für Operations Research*, **24**: B259–B273.

Ng, K.-C. and B. Abramson (1990): "Uncertainty Management in Expert Systems." In: *IEEE Expert*, April: 29–48.

Parsons, L. J. and R. L. Schultz (1978): "Marketing Models and Econometric Research." 2nd Edition, North Holland, New York.

Pedrycz, W. (1983): "Numerical and Applicational Aspects of Fuzzy Relational Equations." In: *Fuzzy Sets and Systems*, **11**: 1–18.

Pedrycz, W. (1985): "Applications of Fuzzy Relational Equations for Methods of Reasoning in the Presence of Fuzzy Data." In: *Fuzzy Sets and Systems*, **16**: 163–175.

Rao, A. G. and M. Yamada (1988): "Forecasting with a Repeat Purchase Diffusion Model." In: *Management Science*, **34**: 734–752.

Shortliffe, E. H. (1984): "Details of the Consultation System." In: R. G. Buchanan and E. H. Shortliffe (eds.): Rule-based Expert Systems, Addison-Wesley, Reading, MA, pp. 78–132.

Spetzler, C. and C.-A. Stael v. Holstein (1975): "Probability Encoding in Decision Analyis." In: *Management Science*, **22**: 340–358.

Sultan, F., J. U. Farley, and D. R. Lehmann (1990): "A Meta-Analysis of Applications of Diffusion Models." In: *Journal of Marketing Research*, **27**: 70–77.

Tellis, G. J. (1988): "The Price-Elasticity of Selective Demand. A Meta-Analysis of Econometric Models of Sales." In: *Journal of Marketing Research*, **25**: 331–341.

Tversky, A. and D. Kahnemann (1975): "Judgement under Uncertainty. Heuristics and Biases." In: D. v. Wendt and C. Vlek (eds.): Utility, Probability and Human Decision Making, Dordrecht, Boston, MA, pp. 141–162.

Vanhonacker, W. R., D. R. Lehman, and F. Sultan (1989): "Combining Related and Sparse Data in Linear Regression Models." Revised Working Paper, INSEAD, Fontainebleau.

Zadeh, L. A. (1965): "Fuzzy Sets." In: *Information and Control*, 338–353.

Part II

The Time Series Approach

Chapter 6

Interpolation of Economic Time Series, with Application to German and Swedish Data

JOHN S. CHIPMAN* AND BEVERLY J. LAPHAM

1 Introduction

It is a common occurrence in empirical work that some time series are available on a monthly basis, some quarterly, and some only on an annual basis, and one wishes to construct a higher-frequency series from a low-frequency one (say quarterly from annual) given information on related time series that are available at a higher frequency. Two examples of this are: (1) stock variables like plant and equipment, information on which may be available only at particular times of the year (say the first quarter); and (2) flow variables such as national income or price level, which may be available only for the entire year. In the case of a stock variable, one would want the interpolated series to agree with the actual series at the point in time when information on the latter is available. In the case of a flow variable, one would want the sum (or average, as the case may be) of the interpolated (say quarterly) series to agree with the the actual annual series.

---

*Work supported by NSF grant SES86-07652. Section 2.2 of this paper builds on some earlier formulations by Andreas Hornstein. We wish to thank Professor Michael Powell for furnishing us with the source code for his nonlinear-programming algorithm (Powell, 1989), and Wayne Fuller, Baldev Raj, and Thomas Url for stimulating comments.

In the econometric literature (e.g., Chow and Lin [1971], Fernández (1981)) these two types of problems have been described as that of "interpolation" (for stock variables) and "distribution" (for flow variables). In the present paper we consider both of these problems to be interpolation problems, with different consistency requirements. The general formulation we present covers both as special cases. As is not surprising, government agencies responsible for issuing published data frequently resort to various methods of interpolation. In the statistical literature these procedures have come to be described as "benchmarking" (cf., e.g., Hillmer and Trbelsi [1987]).

In interpolation thus broadly understood, one seeks to predict unobserved high-frequency values of a series using observed low-frequency values for the series, observed high-frequency values for related series, and trend terms. Such methods are designed, of course, for situations in which data on the high-frequency values are unavailable; however, in order to evaluate the methods of interpolation themselves, they must be confronted either with data sets which include the high-frequency observations, or with simulated data sets if actual high-frequency data are not available. Both types of evaluation are carried out in this paper. Moreover, we address ourselves in particular to two issues: "discontinuities" that are alleged to occur in interpolated values from one low-frequency period to the next, and negatives that may occur in interpolated values.

A number of authors including Friedman [1962], Chow and Lin [1971], Fernández [1981], Litterman [1983], and Harvey and Pierse [1984] have examined methods of interpolation, especially "distribution". Some of their methods are discussed in detail in Section 2. Fernández and Litterman criticize the approach employed by Chow and Lin, arguing that their method of interpolation under the assumption of independent error terms introduces artificial "discontinuities" in the interpolated values. Specifically, it is argued that the Chow-Lin procedure will tend to produce interpolated series which exhibit smoothness over each year but tend to have jumps from one year to the next. No analytic arguments or empirical evidence have been provided to support this contention, however. In Section 2.2 we show that such "discontinuities" do not appear to arise in a variety of data sets, and in only one data set that we have examined may be a problem. We show this for five German data sets for which high-frequency data are available, namely import price indices, export price indices, hours worked, employment, and volume of sales.

All of the methods explored by the above authors allow for negatives in the interpolated values. Using their interpolation methods we sought to predict quarterly building (resp. equipment) investment at the industry level using annual observed building (resp. equipment) investment for the industry; the quarterly observed index of industrial production for the industry; quarterly observed aggregate building (resp. equipment) investment; and spline trend terms. For Germany, the data consisted of 36 industries for building investment and 36 industries for equipment investment from 1962–1984. For Sweden, the data consisted of 58 industries for building investment and 53 industries for equipment investment from 1970–1985. Under the assumption of independent error terms, three industries and one industry exhibited negative interpolated values for West German building and equipment investment respectively. For Sweden, eleven industries and four industries exhibited negative interpolated values for building and equipment investment respectively. Negatives also arose under the assumption of random-walk error terms. These negatives suggest the need for an interpolation method which guarantees nonnegative interpolated values. Such a method is introduced in Section 2.3. Since in this case no quarterly data are available at the industry level, the method is tested against simulated data. It is applied to the German and Swedish investment data.

Finally, a Bayesian approach is developed in Section 3. The time periods are divided into the last four years and the preceding years. Ordinary least-squares estimates computed over the first time period are used to obtain prior means and covariances of the regression coefficients to be used in the second, shorter, period. These are found to provide distinctly superior interpolations over the second period than either of the standard methods applied to the whole period.

The remainder of the paper is organized as follows. Section 2.1 describes the standard interpolation methods introduced by the above authors; Section 2.2 explores the issue of discontinuity by examining various German data sets; and Section 2.3 develops a method for guaranteeing nonnegatives in the interpolated series. Section 3 discusses a Bayesian approach, and Section 4 concludes.

2 Methodology

2.1 Standard Interpolation Methods

Below, we describe some of the standard techniques in the literature for interpolation and distribution and discuss issues of discontinuities and negatives in interpolated values.

Define the following notation:

y : a $T \times 1$ vector of unobserved high-frequency values of the series of interest.

z : an $S \times 1$ vector of observed low-frequency values of the series of interest. ($S = T/4$ if interpolating quarterly data from annual, $S = T/12$ if interpolating monthly data from annual, and $S = T/3$ if interpolating monthly data from quarterly.)

X : a $T \times K$ matrix of K observed high-frequency related series, including trend terms.

G : an $S \times T$ grouping (aggregation) matrix,[1] i.e. $z = Gy$.

Generally, letting $n = T/S$, G can be written as:

$$G = I_S \otimes \phi \tag{6.1}$$

where

I_S is the $S \times S$ identity matrix,
ϕ is a $1 \times n$ vector.

Typical examples of vectors ϕ are the following:

1. The coordinate vector δ_r with 1 in rth place and zeros elsewhere. This would correspond to the case of a stock variable for which data are available only in the rth quarter of each year. The interpolated series $\hat{y}_t$ should agree with the observed series z_s at times $t = r + (s-1)n$ for $s = 1, 2, \ldots, S$.

2. The row vector ι_n of n ones. This would correspond to the case of a flow variable like national income. The interpolated series $\hat{y}_t$ should sum to the corresponding observed series z_s according to $\sum_{t=ns-n+1}^{ns} \hat{y}_t = z_s$ for $s = 1, 2, \ldots, S$.

3. The row vector whose elements are all equal to $1/n$. This would correspond to a flow variable such as an annual price index. The interpolated series $\hat{y}_t$ should average to the corresponding observed series z_s according to $\sum_{t=ns-n+1}^{ns} \hat{y}_t / n = z_s$ for $s = 1, 2, \ldots, S$.

---

[1]Grouping matrices were introduced by Prais and Aitchison [1953] in the study of regression with grouped observations. See also Haitovsky [1973].

Assume the relationship between the series of interest and the related series is given by:

$$y = X\beta + \varepsilon \tag{6.2}$$

where β is $K \times 1$ and ε is $T \times 1$ with $E(\varepsilon) = 0$ and $E(\varepsilon\varepsilon') = V$. Then we write the following:

$$z = Gy = GX\beta + G\varepsilon. \tag{6.3}$$

We seek an interpolator, $\tilde{y}$, of the high-frequency series, y, which satisfies the following criteria:

C1	$\tilde{y} = Az + c$ where A is $T \times S$ and c is $T \times 1$	(linear interpolator)
C2	$E(\tilde{y} - y) = 0$ for all β	(unbiased interpolator)
C3	tr Var $(\tilde{y} - y)$ is minimized	(best interpolator)
C4	$z = G\tilde{y}$	(time aggregation)

Equations (6.3) and (C1) imply the following:

$$E(\tilde{y} - y) = E(Az + c - y) = E(AG - I)y + c = (AG - I)X\beta + c.$$

Therefore, the unbiasedness requirement, (C2), implies the following:

$$(AG - I)X = 0 \quad \text{and} \quad c = 0. \tag{6.4}$$

Equations (6.2)–(6.4) and (C1) imply

$$\tilde{y} - y = Az - y = (AG - I)y = (AG - I)(X\beta + \varepsilon) = (AG - I)\varepsilon.$$

Therefore, we write

$$\text{Var}(\tilde{y} - y) = E[(AG - I)\varepsilon\varepsilon'(AG - I)'] = (AG - I)V(AG - I)'. \tag{6.5}$$

The above equations imply that we can find a best, linear, unbiased (BLU) interpolator of y, $\tilde{y} = Az$, by solving the following constrained-minimization problem:

$\mathcal{P}1$ $\qquad \min\limits_{A} \text{tr}[(AG - I)V(AG - I)'] \quad \text{subject to } (AG - I)X = 0.$

As will be demonstrated below, the solution to this problem also satisfies the aggregation constraint (C4), so this does not need to be imposed as an additional constraint.

Chow and Lin [1971] demonstrate that the solution to $\mathcal{P}1$ implies the following best, linear, unbiased (BLU) interpolator of the high-frequency series:

$$\tilde{y} = X\tilde{\beta} + VG'(GVG')^{-1}(z - GX\tilde{\beta}) \tag{6.6}$$

where

$$\tilde{\beta} = [X'G'(GVG')^{-1}GX]^{-1}X'G'(GVG')^{-1}z. \qquad (6.7)$$

(A formal proof of this is given in Appendix A, Theorem 2.) Note that equation (6.6) implies

$$G\tilde{y} = GX\tilde{\beta} + GVG'(GVG')^{-1}(z - GX\tilde{\beta}) = z,$$

demonstrating that the aggregation requirement, (C4), is satisfied by this interpolator.

If the error terms in equation (6.2) are assumed to be uncorrelated and homoskedastic ($V = \sigma^2 I$),[2] then the interpolator in equation (6.6) becomes

$$\tilde{y} = X[X'G'(GG')^{-1}GX]^{-1}X'G'(GG')^{-1}z + G'(GG')^{-1}(z - GX\tilde{\beta}) \qquad (6.8)$$

where

$$\tilde{\beta} = [X'G'(GG')^{-1}GX]^{-1}X'G'(GG')^{-1}z.$$

Since from (6.1) we have

$$G^{\dagger} = G'(GG')^{-1} = I_S \otimes \phi'(\phi\phi')^{-1} = I_S \otimes \phi^{\dagger} \qquad (6.9)$$

(where $G^{\dagger}$ and $\phi^{\dagger}$ denote the generalized inverses of G and ϕ respectively), if the vector ϕ has equal components (as in examples 2 and 3 above), then under the iid specification each low-frequency residual (i.e., each element of $(z - GX\tilde{\beta})$) is distributed evenly among the n sub-periods.[3] For this reason Fernández [1981] and Litterman [1983] are concerned with the possibility that if error terms are assumed to be independent, the interpolator in equation (6.8) may introduce a "discontinuity" between the last period of one year (if interpolating quarterly or monthly data from annual) and the first period of the next year when the annual residuals are not uniform.

---

[2]We shall refer to this for convenience as the iid (independently identically distributed) case, although only the weaker assumption is used. It should be noted further that the assumption can be weakened to the condition that there exist a matrix W such that $VX = XW$; see, e.g., Kruskal [1968].

[3]$GG' = I, nI$, and I/n in cases 1, 2, and 3 respectively, hence $G^{\dagger} = G', G'/n$, and nG' respectively. If ϕ has unequal components (as in example 1) it will no longer be the case that $G^{\dagger}$ is proportional to G'. Another example of this is the German annual agricultural price index, which is a weighted but not simple average of the monthly indices. Thus, the even spreading of the residuals among sub-periods simply reflects the equal weighting of these sub-periods in the aggregation process. Either autocorrelation or heteroskedasticity would also give rise to uneven spreading of the residuals.

Fernández, then, to correct for this apparent problem, proposes random-walk error terms

$$\varepsilon_t = \varepsilon_{t-1} + \eta_t$$

where

$$E\eta_t = 0, \quad E\eta_t^2 = \sigma^2, \quad E\eta_t\eta_{t'} = 0\ (t' \neq t) \quad (t = 1, \ldots, T)$$

and derives the following interpolator under that assumption:

$$\breve{y} = X\breve{\beta} + (D'D)^{-1}G'\left(G(D'D)^{-1}G'\right)^{-1}(z - GX\breve{\beta}) \tag{6.10}$$

where

$$\breve{\beta} = \left[X'G'\left(G(D'D)^{-1}G'\right)^{-1}GX\right]^{-1}X'G'\left(G(D'D)^{-1}G'\right)^{-1}z$$

and D denotes the $T \times T$ matrix

$$\begin{bmatrix} 1 & 0 & 0 & \ldots & 0 & 0 \\ -1 & 1 & 0 & \ldots & 0 & 0 \\ 0 & -1 & 1 & \ldots & 0 & 0 \\ & & & \ddots & & \\ 0 & -1 & 1 & \ldots & 1 & 0 \\ 0 & 0 & 0 & \ldots & -1 & 1 \end{bmatrix}.$$

He argues that this interpolator will not introduce the discontinuities apparent in the interpolators given by equation (6.8). Litterman [1983], using similar techniques, explores first-order Markov error terms.

We seek to explore two issues related to these interpolators in the remainder of this section. First, we analyze the discontinuity issue by examining the Chow-Lin interpolator given by (6.8) and the Fernández interpolator given by (6.10) as applied to various types of German data for which high-frequency observations are available. It is concluded that discontinuities do not seem to arise in the interpolators given by equation (6.8) under the assumption of independent error terms, and that in one case where they may be a problem, they are not completely surmounted by the random-walk approach. Secondly, we address the problem of negatives in interpolated values which could arise in both of the above approaches. We introduce a method for guaranteeing nonnegative interpolated values, test the method on simulated data, and apply it to German and Swedish investment data.

2.2 "Discontinuities"

We explore the issue of whether the assumption of independent error terms and interpolators of the form given by equation (6.8) give rise to "discontinuities" from the last quarter of one year to the first quarter of the next in five sets of German data. Data sets were selected for which observed high-frequency values are available for comparison with the interpolated series.

2.2.1 Data

The five data sets are as follows:

(1) German Import Price Indices

1970–1982, base year 1976, 35 industries.

Related series are import unit values[4] for each industry, the index of industrial production for the industry, and a spline trend term with three evenly spaced knots.

(2) German Export Price Indices

1970–1982, base year 1976, 35 industries.

Related series are export unit values for each industry, the index of industrial production for the industry, and a spline trend term with three evenly spaced knots.

(3) German Hours Worked (Thousands of man-hours per month)

1962–1985, 30 industries.

Related series are employment for each industry, the index of industrial production for the industry, and a spline trend term with three evenly spaced knots.

(4) German Employment (Number of employees)

1962–1985, 30 industries.

---

[4]Most econometric studies in the field of international trade employ unit values because of the unavailability of true import and export price indices. For discussion of the lack of reliability of unit values see Kravis and Lipsey [1974].

Related series are hours worked for each industry, the index of industrial production for the industry, and a spline trend term with three evenly spaced knots.

(5) German Sales Volumes (Quantity Indices)

1976–1985, 1980 prices, 34 industries.

Related series are the index of industrial production for each industry and a spline trend term with three evenly spaced knots.

In all of the above data sets, the series in question are annual and the related series are quarterly; and quarterly series are also available which can be compared with the interpolated series. In the cases of data sets (1) and (2) the annual series is the simple average of the quarterly series; in the remaining cases it is the sum.

It is implicit in the analysis of the previous subsection that the same model holds for annual, quarterly, or monthly data, in particular, that the same regression coefficients apply in all three models. Given that these are in fact available for most of the above data sets, this hypothesis can be tested. Figures 6.1–6.5 present the relationships among the regression coefficients at these different frequencies. In figures 6.1–6.4 the x-axis measures the respective quarterly regression coefficient relative to the annual one, and the y-axis the corresponding monthly regression coefficient relative to the annual one. In a few of the graphs, outliers (indicated in parentheses) have been eliminated since they distorted the scale.[5] Most of the points are distributed along the 45°-line,[6] indicating that that the monthly and quarterly regression coefficients agree quite well; furthermore, most of them are concentrated near the point $(1, 1)$, indicating that they agree fairly well with the annual coefficients. In the case of sales volumes (figure 6.5), monthly data while available were not at our disposal at the time the computations were carried out, hence the x-axis measures annual coefficients and the y-axis quarterly coefficients; here, the agreement is not so good.

In the subsequent analysis, three methods are employed to examine the discontinuity issue using these sets of data. In what follows, let J be the

[5]These are as follows: In figure 6.1, the pair of coefficients for quarterly and monthly machinery import unit values (both relative to annual) was $(323.71, 251.59)$; in figure 6.2, the pair for quarrying products unit values was $(-15.03, -17.85)$ and that for the spline term for iron, sheet and other metal goods was $(-13.85, -9.14)$; and in figure 6.4 the pair for the index of industrial production for printed products was $(18.23, 0.30)$.

[6]Note that since the x- and y-axes are differently scaled, the angle in the diagrams is distorted.

number of industries in the data set and let y^j be the value of the series for the jth industry. Each method is discussed in turn.

2.2.2 Metric Statistics

We define a metric, $\Delta(\alpha)(\hat{y}^j, y^j)$, which is a measure of the distance between the interpolated series, $\hat{y}^j$, and the actual series, y^j, for the jth industry, as follows, for $\alpha \in [0,1]$:[7]

$$\Delta(\alpha)(\hat{y}^j, y^j) = \frac{1}{T}\left[\sum_{t=1}^{T}(y_t^j - \hat{y}_t^j)^2\right]^{1/2} + \frac{\alpha}{T-1}\left[\sum_{t=2}^{T}[(y_t^j - y_{t-1}^j) - (\hat{y}_t^j - \hat{y}_{t-1}^j)]^2\right]^{1/2}. \quad (6.11)$$

Means, standard deviations and the range of the metrics for $\alpha = 0$ and $\alpha = 1$ for each data set for the interpolator given by equation (6.8) under the assumption of independent error terms and for the interpolator given by equation (6.10) under the assumption of random-walk error terms are presented in tables 6.1–6.5.

Table 6.1:
Metric Statistics for Import Price Indices

	IID Error Terms		Random-Walk Error Terms	
	$\Delta(0)$	$\Delta(1)$	$\Delta(0)$	$\Delta(1)$
Mean	3.46	5.11	3.70	5.57
Standard deviation	2.52	3.75	3.46	5.27
Range	1.21	1.61	1.10	1.61
	12.78	16.25	19.57	29.32

Table 6.2:
Metric Statistics for Export Price Indices

	IID Error Terms		Random-Walk Error Terms	
	$\Delta(0)$	$\Delta(1)$	$\Delta(0)$	$\Delta(1)$
Mean	2.62	3.94	2.59	3.94
Standard deviation	1.53	2.26	2.17	3.31
Range	0.81	1.22	0.79	1.09
	7.36	10.27	12.21	18.50

[7]This measure was suggested by some criteria used in Whittaker and Robinson [1924]. It also corresponds to the standard metric employed in the calculus of variations (cf., e.g., Akhiezer [1962]).

Table 6.3:
Metric Statistics for Hours Worked

	IID Error Terms		Random-Walk Error Terms	
	$\Delta(0)$	$\Delta(1)$	$\Delta(0)$	$\Delta(1)$
Mean	413.01	679.20	405.47	671.19
Standard deviation	959.62	1624.72	989.57	1666.13
Range	14.30	23.33	12.74	19.46
	5231.55	8814.94	5258.63	8386.74

Table 6.4:
Metric Statistics for Employment

	IID Error Terms		Random-Walk Error Terms	
	$\Delta(0)$	$\Delta(1)$	$\Delta(0)$	$\Delta(1)$
Mean	2051.29	3293.07	1758.76	2771.70
Standard deviation	2460.78	4095.29	2142.43	3410.43
Range	124.88	186.62	109.40	167.60
	11634.9	19352.24	10177.16	15836.56

Table 6.5:
Metric Statistics for Sales Volumes

	IID Error Terms		Random-Walk Error Terms	
	$\Delta(0)$	$\Delta(1)$	$\Delta(0)$	$\Delta(1)$
Mean	0.95	1.43	0.83	1.24
Standard deviation	0.99	1.53	0.98	1.52
Range	0.22	0.36	0.22	0.36
	5.42	8.39	5.45	8.48

These tables illustrate that neither interpolator performs markedly better than the other for these data sets. Recall that if discontinuities were being introduced erroneously, then the random-walk interpolators should perform better than the iid interpolators. This does not appear to be the case.

2.2.3 Quarterly Differences

For each data set and each year, $s = 1, \ldots, S$, in the data set, let the industry average of the absolute value of the quarterly differences be given by

$$\delta_i^s = \frac{1}{J}\sum_{j=1}^{J}\left|y_{4s-3+i}^j - y_{4s-4+i}^j\right| \qquad \text{for } i = 1,2,3; \quad s = 1,2,\ldots,S$$

$$\delta_4^s = \frac{1}{J}\sum_{j=1}^{J}\left|y_{4s+1}^j - y_{4s}^j\right| \qquad \text{for } s = 1,2,\ldots,S-1.$$

Thus, for $i = 1,2,3$, δ_i^s measures the industry average of the absolute value of the differences between the observation in quarter $i+1$ and the observation in quarter i in year s. δ_4^s measures the average differences between the first quarter of year $s+1$ and the fourth quarter of year s.

Let the averages over time of these quarterly-difference statistics be given as follows:

$$\bar{\delta}_i = \frac{1}{S}\sum_{s=1}^{S}\delta_i^s \quad \text{for } i = 1,2,3$$

$$\bar{\delta}_4 = \frac{1}{S-1}\sum_{s=1}^{S-1}\delta_4^s.$$

Tables 6.6–6.10 report these averages for the actual data and the interpolators given by equations (6.8) and (6.10).

Table 6.6:
Average Absolute Quarterly Differences for Import Price Indices

	$\bar{\delta}_1$	$\bar{\delta}_2$	$\bar{\delta}_3$	$\bar{\delta}_4$
Data	20.02	19.90	19.70	33.63
IID	30.16	32.00	33.61	49.01
Random Walk	33.49	42.93	38.10	42.15

Table 6.7:
Average Absolute Quarterly Differences for Export Price Indices

	$\bar{\delta}_1$	$\bar{\delta}_2$	$\bar{\delta}_3$	$\bar{\delta}_4$
Data	16.78	13.49	12.56	27.04
IID	26.55	29.54	25.16	35.22
Random Walk	25.01	32.72	26.15	32.26

Table 6.8:
Average Absolute Quarterly Differences for Hours Worked

	$\bar{\delta}_1$	$\bar{\delta}_2$	$\bar{\delta}_3$	$\bar{\delta}_4$
Data	2980.49	3719.97	4836.51	3703.97
IID	2364.53	3085.48	3093.92	4015.64
Random Walk	3249.72	4087.29	4253.68	4254.46

Table 6.9:
Average Absolute Quarterly Differences for Employment

	$\bar{\delta}_1$	$\bar{\delta}_2$	$\bar{\delta}_3$	$\bar{\delta}_4$
Data	10056.25	8589.91	8790.95	17175.44
IID	21803.16	20827.19	28034.25	30273.00
Random Walk	19649.51	16975.41	21667.09	23126.47

Table 6.10:
Average Absolute Quarterly Differences for Sales Volumes

	$\bar{\delta}_1$	$\bar{\delta}_2$	$\bar{\delta}_3$	$\bar{\delta}_4$
Data	10.95	10.39	14.34	13.15
IID	6.89	8.87	10.78	8.54
Random Walk	7.12	8.22	10.06	7.55

If the estimation technique employed under iid error terms was introducing an artificial discontinuity between the last quarter of one year and the first quarter of the next, then one would expect $\bar{\delta}_4$ to be larger relative to $\bar{\delta}_i$ for $i = 1, 2, 3$ in the iid estimates than in the actual data. The above tables do not seem to indicate that this is the case. A more rigorous test of this assertion is presented below.

Define the following ratio statistics:

$$\begin{aligned} R_i^s &= \delta_4^s / \delta_i^s && \text{for } i = 1,2,3;\ s = 1,\ldots,S-1 \\ \bar{R}_i &= \frac{1}{S-1}\sum_{s=1}^{S-1} R_i^s && \text{for } i = 1,2,3. \end{aligned}$$

For each $i = 1, 2, 3$, let

$\bar{R}_i$ be the above statistic $\bar{R}_i$ computed from the actual data;

$\bar{\tilde{R}}_i$ be the statistic $\bar{R}_i$ computed from the interpolated data $\tilde{y}_t^j$ under iid error terms;

$\bar{\check{R}}_i$ be the statistic $\bar{R}_i$ computed from the interpolated data $\check{y}_t^j$ under random-walk error terms.

Consider the following one-tailed tests, for $i = 1, 2, 3$:

$$\begin{array}{lll} \text{Test } \tilde{T}_i: & H_0: \bar{\tilde{R}}_i - \bar{R}_i = 0; & H_a: \bar{\tilde{R}}_i - \bar{R}_i > 0; \\ \text{Test } \check{T}_i: & H_0: \bar{\check{R}}_i - \bar{R}_i = 0; & H_a: \bar{\check{R}}_i - \bar{R}_i > 0. \end{array}$$

If the null hypothesis is not rejected, we may conclude that discontinuities are not being introduced by the interpolation methods.

Let $\tilde{t}_i, \check{t}_i$ be the t-statistics associated with the tests $\tilde{T}_i, \check{T}_i$, i.e. (cf., e.g., Fisher, 1958, p. 122).[8] For example,

$$\tilde{t}_i = \frac{\bar{\tilde{R}}_i - \bar{R}_i}{\tilde{\nu}} \sqrt{\frac{S-1}{2}}$$

where

$$\tilde{\nu}^2 = \frac{\sum_{s=1}^{S-1} (\tilde{R}_i^s - \bar{\tilde{R}}_i)^2 + \sum_{s=1}^{S-1} (R_i^s - \bar{R}_i)^2}{2(S-2)},$$

and similarly for the test $\check{T}_i$. The values of these t-statistics and the degrees of freedom (DOF) for each data set are reported in tables 6.11–6.12.

[8] If, for example, $\tilde{R}_i^s$ and R_i^s are independently normally distributed with common variance ν^2 (where $\tilde{R}_i^s = \tilde{\delta}_4^s / \tilde{\delta}_i^s$ and the $\tilde{\delta}_i^s$ are defined in the obvious way in terms of the $\tilde{y}_t^j$), then $\tilde{t}_i$ has the t distribution with $2(S-2)$ degrees of freedom. A sufficient condition for $\tilde{R}_i^s$ and R_i^s to be independently distributed would be that $\tilde{y}$ and y be independently distributed. Unfortunately, however, this is not so in the present case, since the covariance between $\tilde{y}$ and y is in fact $\sigma^2 M$, where

$$M = \{I + [I - G'(GG')^{-1}G]X[X'G'(GG')^{-1}GX]^{-1}X'\}G'(GG')^{-1}G.$$

Furthermore, the degrees of freedom are not sufficiently high to ensure that $\bar{\tilde{R}}_i^s$ and $\bar{R}_i^s$ are approximately normally distributed. Thus, we are implicitly assuming that the t-test is sufficiently robust against departures from assumptions that some meaning can be attached to our results in the absence of an exact test, which would probably be difficult to devise in the present case.

Table 6.11:
t-statistics for IID Error Terms

Data	DOF	99%	$\tilde{t}_1$	$\tilde{t}_2$	$\tilde{t}_3$
Import Prices	22	2.508	-8.83	-6.77	-12.14
Export Prices	22	2.508	-20.44	-33.98	-42.91
Hours Worked	44	2.414	46.28	30.90	61.92
Employment	44	2.414	-36.37	-59.01	-66.86
Sales Volumes	16	2.584	3.71	-25.14	-18.90

Table 6.12:
t-statistics for Random-Walk Error Terms

Data	DOF	99%	$\check{t}_1$	$\check{t}_2$	$\check{t}_3$
Import Prices	22	2.508	-22.54	-29.48	-33.93
Export Prices	22	2.508	-22.11	-44.20	-57.62
Hours Worked	44	2.414	-3.80	-8.61	24.14
Employment	44	2.414	-58.15	-65.68	-66.85
Sales Volumes	16	2.584	-10.09	-29.33	-26.25

Table 6.11 indicates that under the iid estimates, the null hypothesis cannot be rejected at the 99% level for all data sets except hours worked. Table 6.12 indicates that under the random-walk estimates, the null hypothesis cannot be rejected at the 99% level for all data sets except hours worked in test $\check{T}_3$. Therefore, in four out of five of the data sets examined, there is no evidence of "discontinuities" in the iid estimates; indeed, there is some evidence that interpolation introduces *less* inter-year variation than intra-year variation relative to actual values. Moreover, in the hours-worked data, the random-walk estimates do not completely correct for any discontinuities introduced by the iid estimates.

We may also test jointly for the equality of the three pairs of means in the above two cases. Thus, consider the following two tests:

$$\text{Test}\,\tilde{f}: \ H_0: \tilde{\bar{\bar{R}}}_i - \bar{R}_i = 0 \ (i = 1,2,3); \ \ H_a: \tilde{\bar{\bar{R}}}_i - \bar{R}_i \neq 0 \ \text{ for some } i = 1,2,3;$$
$$\text{Test}\,\check{f}: \ H_0: \check{\bar{\bar{R}}}_i - \bar{R}_i = 0 \ (i = 1,2,3); \ \ H_a: \check{\bar{\bar{R}}}_i - \bar{R}_i \neq 0 \ \text{ for some } i = 1,2,3.$$

If the null hypothesis is not rejected, we may conclude that year-to-year discontinuities are not being introduced by the interpolation methods. If it is rejected, we may conclude that they introduce either significantly more or significantly less intra-year than inter-year variations.

Let $\tilde{F}$, $\check{F}$ be the F-statistics associated with the tests $\tilde{f}$, $\check{f}$, with 3 and $6(S-2)$ degrees of freedom. For example,

$$\tilde{F} = \frac{\sum_{i=1}^{3}(\tilde{R}_i - R_i)^2}{\tilde{\nu}^2} \frac{S-1}{2}$$

where

$$\tilde{\nu}^2 = \frac{\sum_{i=1}^{3}\left(\sum_{s=1}^{S-1}(\tilde{R}_i^s - \bar{\tilde{R}}_i^s)^2 + \sum_{s=1}^{S-1}(R_i^s - \bar{R}_i^s)^2\right)}{6(S-2)}.$$

(The same disclaimer regarding lack of independence applies to these tests as to the t-tests—see the previous footnote.) The values of these F-statistics and their numerator and denominator degrees of freedom for each data set are reported in table 6.13.

Table 6.13:
F-statistics for IID and Random-Walk Error Terms

Data	DOF	99%	IID	RW
Import Prices	3, 66	4.093	1.56	15.24
Export Prices	3, 66	4.093	24.41	34.76
Hours	3,132	3.933	13.02	1.22
Employment	3,132	3.933	20.96	24.15
Sales Volumes	3, 48	4.218	10.32	17.69

According to this test, the null hypothesis is not rejected in the iid case for import prices, and in the random-walk case for hours, but rejected in the other cases. But combined with the previous results, this leads us to infer that, if anything, there is greater intra-year than inter-year variation in interpolated values relative to actual values.

2.2.4 Graphics

Finally, graphs of the actual and interpolated data under the two methods are presented for randomly selected industries. These graphs demonstrate how close the interpolated values are to the actual values and illustrate the lack of discontinuities in the method under independent error terms. Figures 6.6–6.10 depict actual values against iid interpolators and against random-walk interpolators for two industries from each sample. The graphs provide no evidence of discontinuities in the iid case.

We can conclude from the German data examined in this study that discontinuities do not appear to be an issue in interpolating these series except, perhaps, for hours worked. A second issue that we now turn to is the possibility of negatives in the interpolated values when the series must be nonnegative.

2.3 Negatives in Interpolation

In this section, a method for predicting high-frequency values for a series when those values must be nonnegative is described. The method is applied to a generated data set which has properties consistent with the investment data discussed in the introduction in which negatives originally arose. The generated data are used for testing the method, and comparisons of the interpolated values with actual values are presented.

2.3.1 Method

The method presented here directly imposes nonnegativity constraints on the minimization problem $\mathcal{P}1$ described in the previous section. This problem can be written as follows:

$$\mathcal{P}2 \qquad \min_{A} \operatorname{tr}[(AG - I)V(AG - I)']$$

subject to

$$AGX = X \tag{6.12}$$

$$GAz = z \tag{6.13}$$

$$Az \geq 0. \tag{6.14}$$

Equation (6.12) guarantees unbiased interpolators, equation (6.13) guarantees the aggregated interpolators equal observed aggregates, and inequality (6.14) guarantees nonnegative interpolators. Note that if we impose nonnegative interpolated values, then the aggregation constraint is no longer automatically satisfied and must be imposed. Generally, except in the case of random-walk error terms, this problem can be written as a series of smaller independent problems as follows:

$$\forall s = 1, \ldots, S \quad \min_{\substack{\{a_{ij}\} \\ i=4s-3,\ldots,4s \\ j=1,\ldots,S}} \sum_{i=4s-1}^{4s} \left[2\sum_{j=1}^{S} a_{ij}^2 - a_{is} \right]$$

subject to

$$\sum_{m=1}^{S} a_{im} \sum_{j=4m-3}^{4m} x_{jq} = x_{iq} \qquad \begin{cases} \forall i = 4s-3,\ldots,4s \\ \forall q = 1,\ldots,K \end{cases}$$

$$\sum_{i=4s-3}^{4s} \sum_{j=1}^{S} a_{ij} z_j = z_s$$

$$\sum_{j=1}^{S} a_{ij} z_j \geq 0 \qquad \forall i = 4s-3,\ldots,4s.$$

(See Theorem 2* in Appendix A.) These problems can be solved using a quadratic-programming routine developed by Powell [1983a, 1983b] on the basis of Goldfarb and Idnani [1983], or by use of a more general constrained-minimization computer routine developed by Powell [1989] for solving general objective functions subject to linear constraints. Owing to the fact that that problem $\mathcal{P}2$ decomposes into S smaller problems, one for each low-frequency period, one may proceed simply as follows: (1) compute the solution to problem $\mathcal{P}1$; (2) if for any $s = 1,\ldots,S$ there are some negative interpolated values, solve the decomposed problem $\mathcal{P}2$ and recompute the parameters for this s.

2.3.2 Simulation

To test the interpolation method, data are generated which exhibit properties consistent with the investment data in which the negative values originally arose. Interpolated values can then be compared with the actual generated series to test the method. Table 6.14 presents relevant statistics for the 19 industries in which negative values appeared in the West German and Swedish investment data. Table 6.15 presents the same statistics for 20 simulated data sets. Details regarding the generation of the simulated data are confined to Appendix B.

Adopt the following notation:

$\bar{z}, \sigma_z, \rho_z$: Mean, standard deviation, and autocorrelation coefficient of the low-frequency observed series (in the investment problem this is annual industry investment).

$\bar{x}_j, \sigma_j, \rho_j$: Mean, standard deviation, and autocorrelation coefficients in the jth high-frequency related series, $j = 1, 2$ (in the investment problem these are quarterly indices of industrial production and quarterly aggregate investment).

ρ_{zj}: Cross-correlation coefficient between the low-frequency series and the jth aggregated high-frequency related series, j= 1, 2.

Table 6.14:
Investment Data (19 observations)

	$\bar{z}$	σ_z	$\bar{x}_1$	σ_1	$\bar{x}_2$	σ_2
Range	17	8	140	21	979	222
	550	348	4599	770	3008	759

	ρ_z	ρ_1	ρ_2	ρ_{z1}	ρ_{z2}
Range	-0.36	-0.13	0.0	-0.57	-0.58
	0.89	0.90	0.65	0.82	0.74

Table 6.15:
Simulated Data (20 observations)

	$\bar{z}$	σ_z	$\bar{x}_1$	σ_1	$\bar{x}_2$	σ_2
Range	161	116	407	198	2606	313
	336	280	452	226	2974	446

	ρ_z	ρ_1	ρ_2	ρ_{z1}	ρ_{z2}
Range	-0.06	0.82	0.26	0.38	-0.34
	0.84	0.87	0.56	0.72	0.46

As these tables demonstrate, the properties of the simulated data are generally consistent with the properties of the investment data, particularly with respect to second moments.

Table 6.16 presents the properties of the metric defined above over the 20 simulated data sets when the data have been interpolated to guarantee nonnegative values under the assumption of independent error terms, using Powell's [1989] nonlinear programming algorithm. This table indicates that the method provides interpolated values which are reasonably close to the simulated data. Imposition of the nonnegativity constraint does not appear to introduce large discrepancies.

Table 6.16:
Metric Statistics for Simulated Data

	IID Error Terms	
	Δ(0)	Δ(1)
Mean	1.90	4.93
Standard deviation	.47	1.25
Range	1.17	3.11
	2.90	7.52

It should be noted that the solution to problem $\mathcal{P}2$ is a boundary solution. The negative interpolated values which arise under the minimization problem $\mathcal{P}1$ (which does not impose nonnegativity constraints) are replaced by zeros in the solution to problem $\mathcal{P}2$ (which does impose nonnegativity constraints) and the low-frequency series is distributed optimally over the remaining subperiods within the period.

This result can be seen by examining figure 6.11. This diagram depicts the simulated data, the interpolated values when nonnegative values are not guaranteed, and the interpolated values when nonnegative values are guaranteed, for two of the twenty simulated data sets. Note that the negative interpolated values are replaced by zeros when nonnegative interpolated values are guaranteed.

2.3.3 Investment Data

The above method was also applied to the German and Swedish investment data in which the negative interpolated values originally arose. The four data sets used were as follows:

(6) German Gross Building Investment

1962–1984, base year 1980, 36 industries. Annual data.

Related series are quarterly industry index of industrial production, quarterly aggregate building investment, and a spline trend term with three evenly spaced knots.

(7) German Gross Equipment Investment

1962–1984, base year 1980, 36 industries. Annual data.

Related series are quarterly industry index of industrial production, quarterly aggregate equipment investment, and a spline trend term with three evenly spaced knots.

(8) Swedish Gross Building Investment

1970–1985, base year 1980. Annual data.

Related series are quarterly industry index of industrial production, quarterly aggregate building investment, and a spline trend term with three evenly spaced knots.

(9) Swedish Gross Equipment Investment

1970–1985, base year 1980. Annual data.

Related series are quarterly industry index of industrial production, quarterly aggregate equipment investment, and a spline trend term with three evenly spaced knots.

2.3.4 Empirical Results

Four out of 72 German industries exhibited negative interpolated values (three in building investment and one in equipment investment) while 15 out of 111 Swedish industries (eleven in building investment and four in equipment investment) exhibited negative values under the assumption of independent error terms. Figures 6.12–6.13 depict data from two randomly selected German industries which exhibited negatives and figures 6.14–6.15 depict data from two randomly selected Swedish industries which exhibited negatives. In each of these diagrams, the low-frequency series (annual industry investment), the related series (quarterly industry index of industrial production and quarterly aggregate investment), and the interpolated series with and without negatives (quarterly industry investment) are presented. These figures also demonstrate that the solution to problem $\mathcal{P}2$ is a boundary solution.

3 Bayesian Analysis

When the available time series are short and unreliable, use of *a priori* information, if available, is indispensible. Ordinarily this would have to come from data in analogous situations, for example from different but somewhat similar countries. How could one verify that procedures using such *a priori* information are satisfactory? A necessary condition, certainly, is that such procedures could be shown to be satisfactory if prior data from the same country were, in fact, available. In this section, therefore, we proceed on this assumption.

Let the time series be partitioned into an early, long, period of length T_1 and a late, short period of length T_2, where $T_1 + T_2 = T$, and let the $T \times K$ observation matrix X and the $T \times 1$ vector y be partitioned into their first T_1 and last T_2 rows according to

$$X = \begin{bmatrix} X_1 \\ X_2 \end{bmatrix}, \quad y = \begin{bmatrix} y_1 \\ y_2 \end{bmatrix}.$$

We shall assume iid errors, i.e., $V = \sigma^2 I$. In the model (6.2) let β be assumed to be a random variable distributed independently of ε, with mean $\bar{\beta}$ and variance U. These prior means and variances may then be estimated by

$$\bar{\beta} = \tilde{\beta} = (X_1'X_1)^{-1}X_1'y_1 \quad \text{and} \quad U = \text{Var}\,\tilde{\beta} = \sigma^2(X_1'X_1)^{-1}. \tag{6.15}$$

In period 2 one wishes to interpolate the vector y_2 using this *a priori* information as well as the related series X_2, when data on only $z_2 = Gy_2$ are available. From formula (6.A18) of Appendix A, this interpolator is

$$\hat{y}_2 = X_2\hat{\beta} + G'(GG')^{-1}(z_2 - GX_2\hat{\beta})$$

where, from (6.A17) and (6.A15), and substituting the prior values (6.15),

$$\hat{\beta} = \bar{\beta} + [X_2'G'(GG')^{-1}GX_2 + X_1'X_1]^{-1}X_2'G'(GG')^{-1}(z_2 - GX_2\bar{\beta}). \tag{6.16}$$

This method has been applied to the last four years of each of our five data sets, using the previous years' data to furnish the *a priori* information. The metric statistics are displayed in tables 6.17–6.21 and compared with those of the iid and random-walk methods (where the regressions are computed over the entire period) during those four years. The Bayes procedure evidently provides superior interpolations. Figures 6.16–6.20 display actual values and values interpolated according to the Bayes procedure for randomly-selected industries in each of the five data sets of subsection 2.2.1; the y-axes in these charts are scaled to permit comparison with figures 6.6–6.10.

Table 6.17:
Metric Statistics for Import Prices (35 Industries)

	IID		Random Walk		Bayes	
	$\Delta(0)$	$\Delta(1)$	$\Delta(0)$	$\Delta(1)$	$\Delta(0)$	$\Delta(1)$
Mean	6.92	17.74	7.94	20.30	5.46	13.58
Standard deviation	5.15	13.69	8.37	21.52	4.85	12.22
Range	1.72	4.78	2.18	5.55	1.77	4.68
	25.55	65.28	43.10	112.04	25.73	63.23

Table 6.18:
Metric Statistics for Export Prices (35 Industries)

	IID		Random Walk		Bayes	
	$\Delta(0)$	$\Delta(1)$	$\Delta(0)$	$\Delta(1)$	$\Delta(0)$	$\Delta(1)$
Mean	5.14	13.42	5.23	13.64	3.49	8.94
Std. deviation	2.77	7.23	4.15	10.66	1.57	4.05
Range	1.42	3.63	1.23	2.96	1.09	2.92
	14.93	38.29	22.35	57.07	7.60	20.28

Table 6.19:
Metric Statistics for Hours Worked (30 Industries)

	IID		Random Walk		Bayes	
	$\Delta(0)$	$\Delta(1)$	$\Delta(0)$	$\Delta(1)$	$\Delta(0)$	$\Delta(1)$
Mean	535.72	1424.16	546.62	1497.49	425.35	1116.70
Std. deviation	776.83	2123.96	1076.96	3019.60	543.60	1454.36
Range	13.92	3.50	6.09	15.04	5.19	12.91
	3891.64	10618.86	5749.53	16056.07	1932.51	5033.88

Table 6.20:
Metric Statistics for Employment (30 Industries)

	IID		Random Walk		Bayes	
	$\Delta(0)$	$\Delta(1)$	$\Delta(0)$	$\Delta(1)$	$\Delta(0)$	$\Delta(1)$
Mean	3644.20	9757.11	2850.18	7550.81	2567.90	6599.45
Std.dev.	4326.86	12187.53	3364.08	9436.08	3205.68	8549.19
Range	116.98	296.51	48.69	119.64	39.98	98.74
	16207.01	46841.31	14851.40	43890.96	13960.79	40095.75

Table 6.21:
Metric Statistics for Sales Volumes (34 Industries)

	IID		Random Walk		Bayes	
	$\Delta(0)$	$\Delta(1)$	$\Delta(0)$	$\Delta(1)$	$\Delta(0)$	$\Delta(1)$
Mean	1.62	4.25	1.41	3.72	1.29	3.40
Std.deviation	1.85	4.98	1.85	4.97	1.79	4.78
Range	0.30	0.82	0.26	0.69	0.25	0.64
	10.39	27.73	10.45	27.69	10.53	28.06

Figure 6.1:
German Import Price Indices

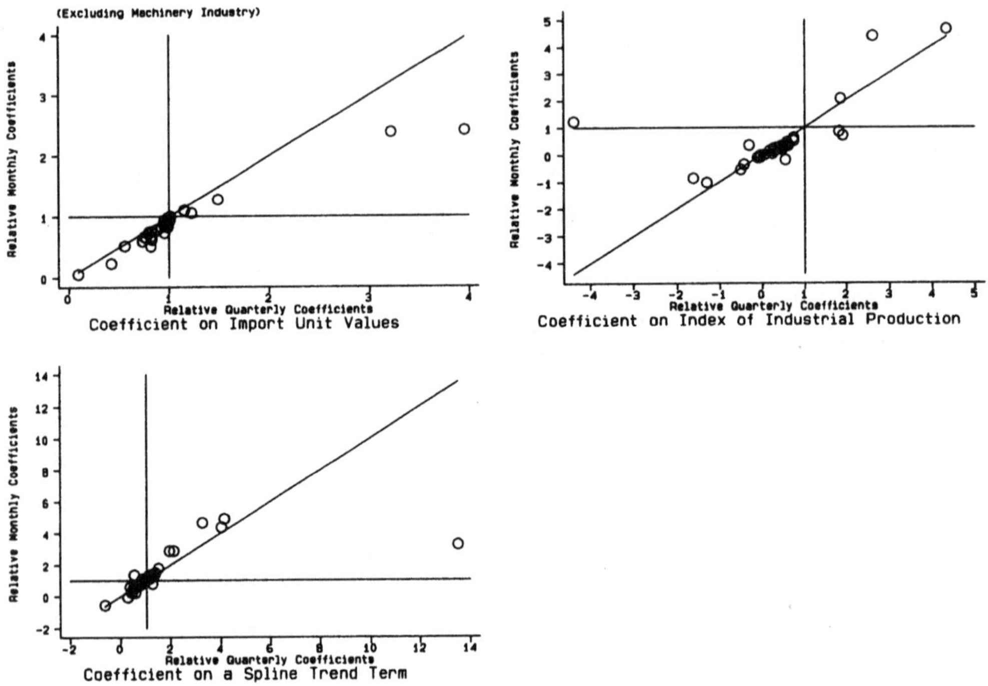

Figure 6.2:
German Export Price Indices

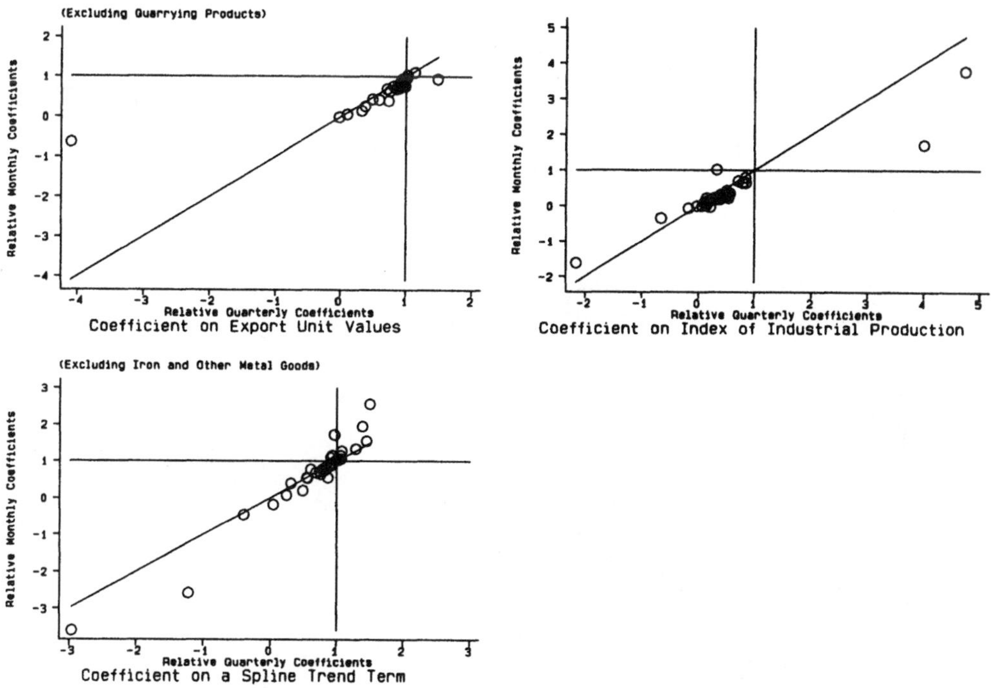

Figure 6.3:
German Hours Worked

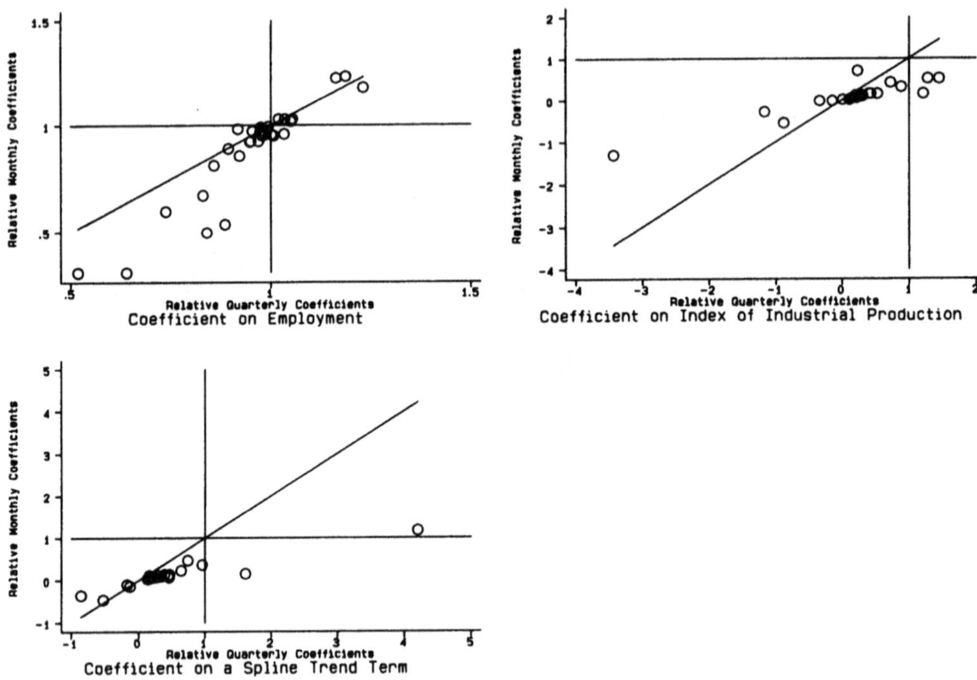

Figure 6.4:
German Employment

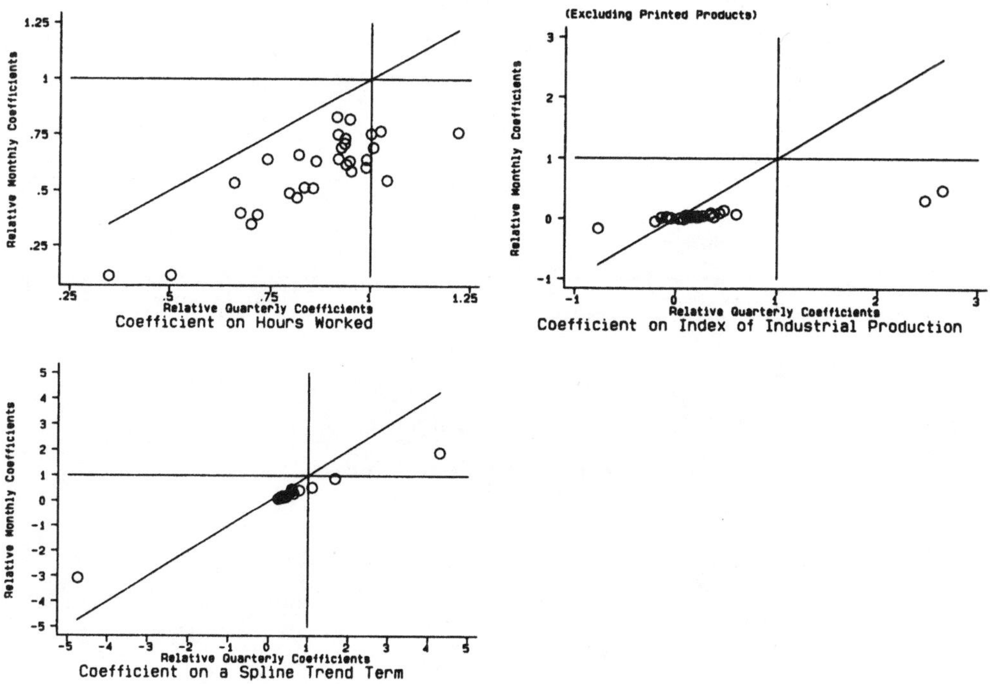

Figure 6.5:
German Sales Volumes

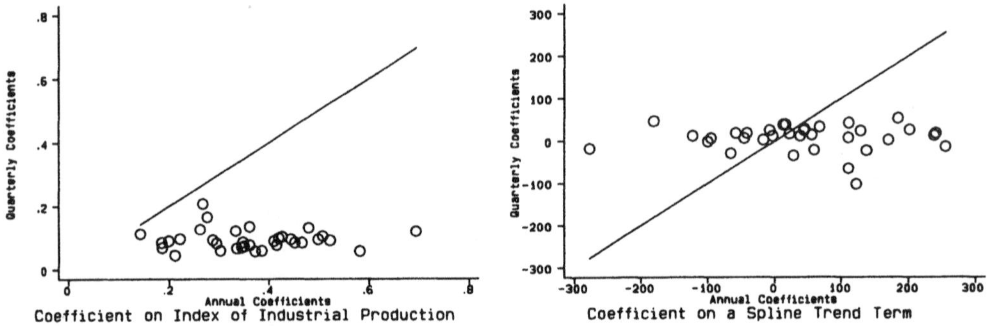

Figure 6.6:
German Import Price Indices

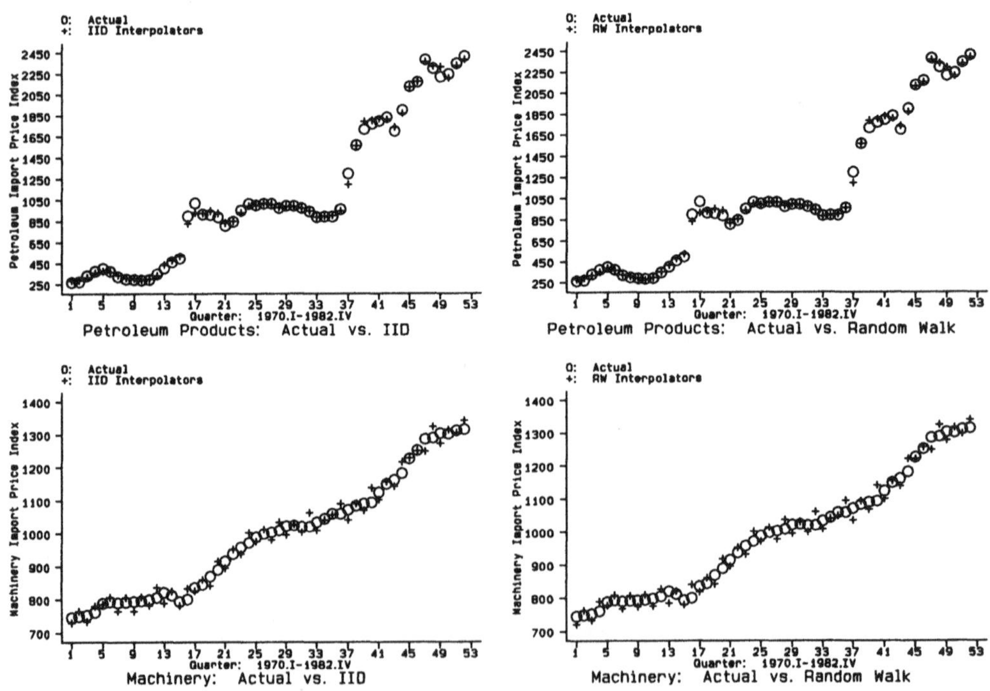

Figure 6.7:
German Export Price Indices

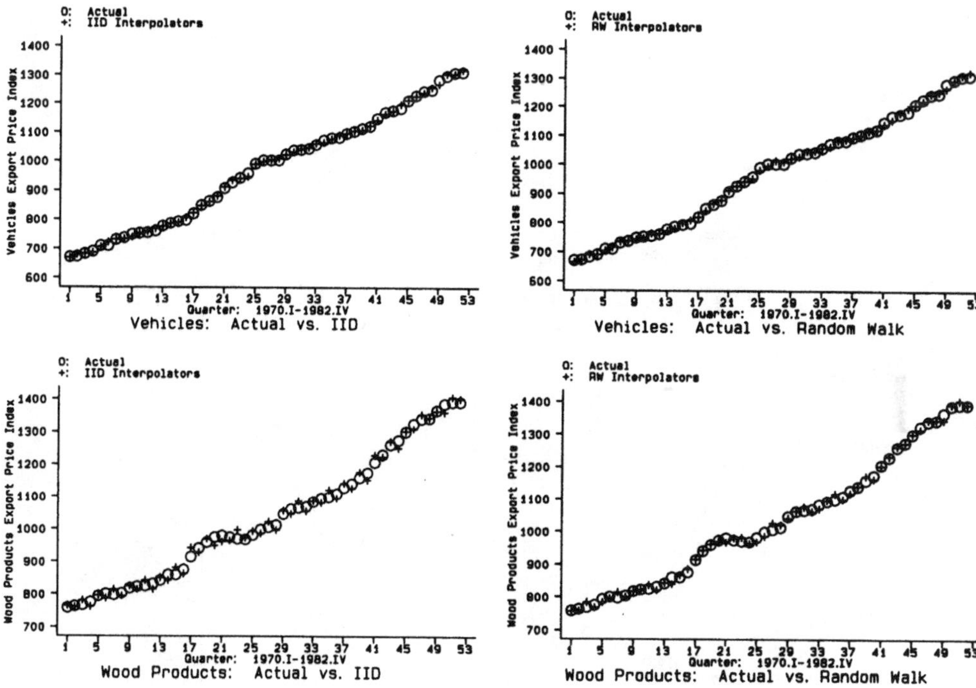

Figure 6.8:
German Hours Worked

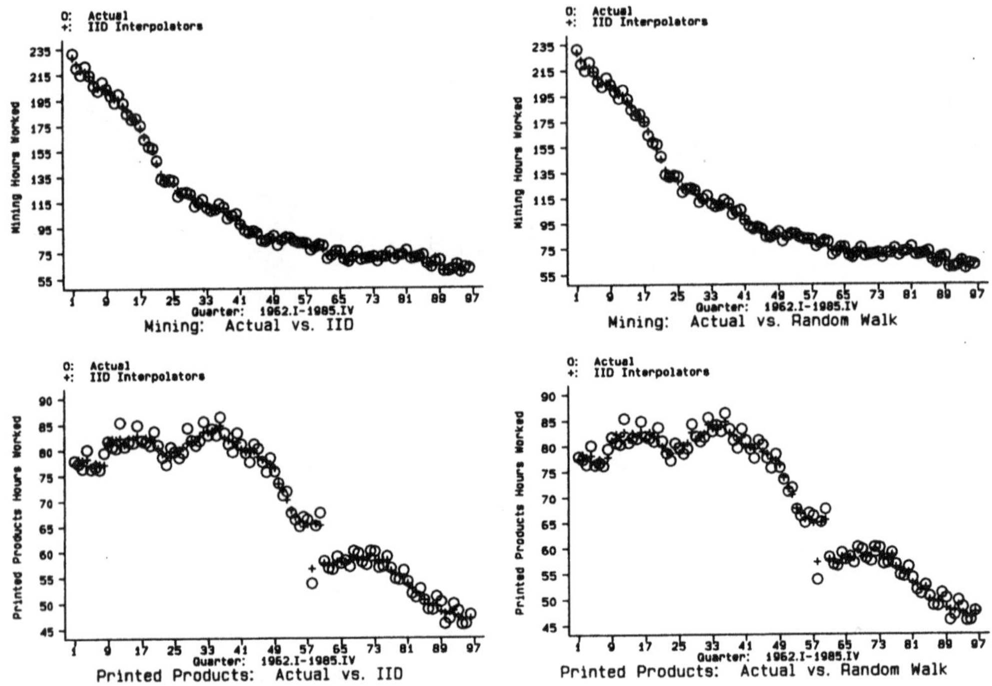

Figure 6.9:
German Employment

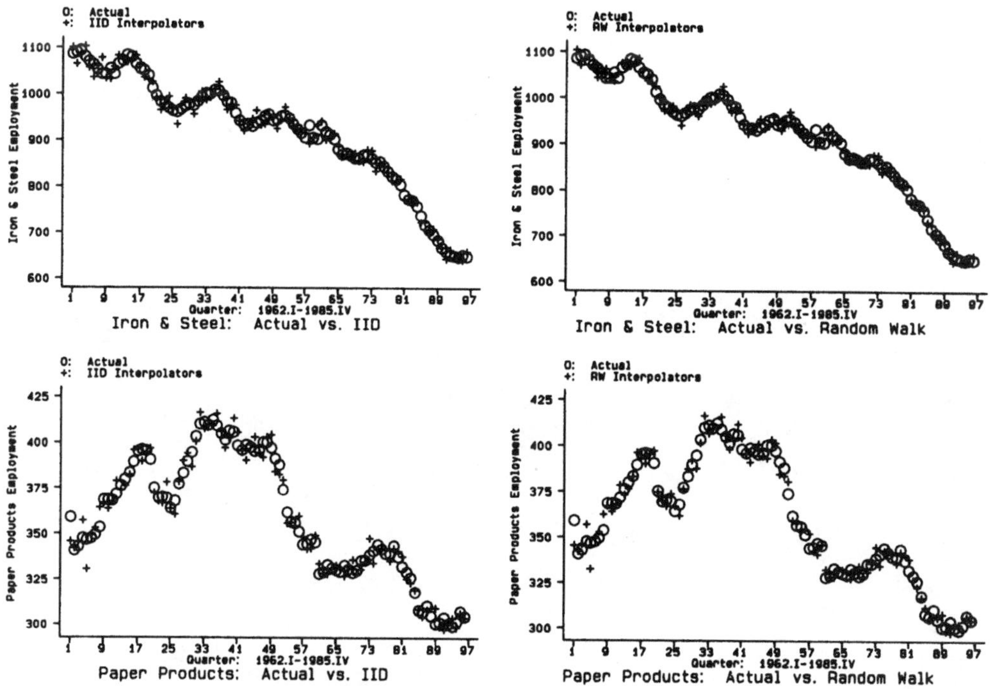

Figure 6.10:
German Sales Volumes

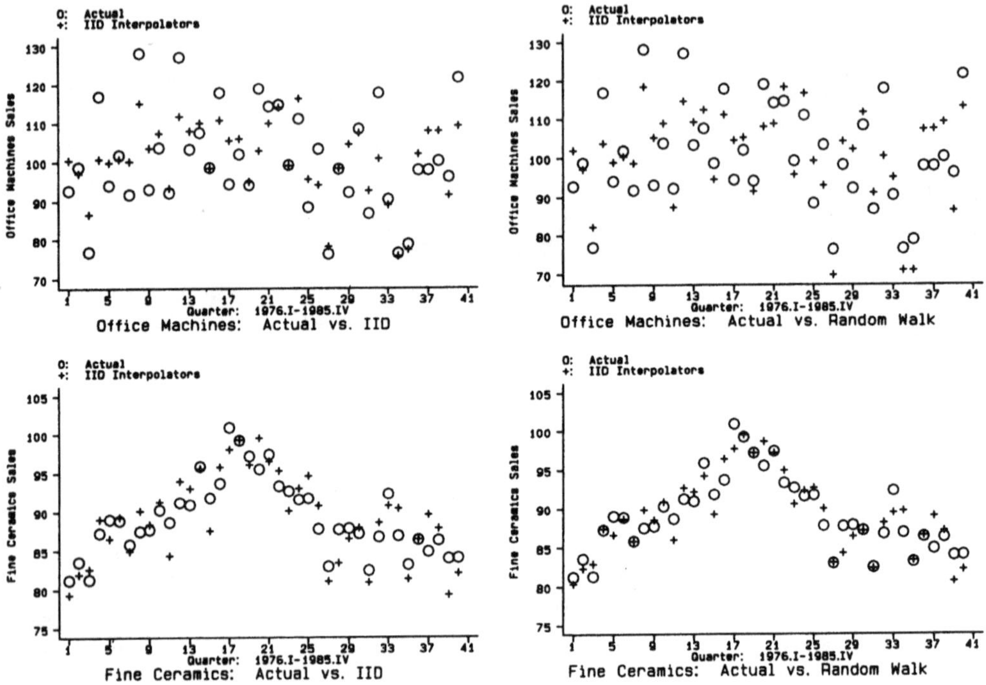

Figure 6.11:
Simulated Data

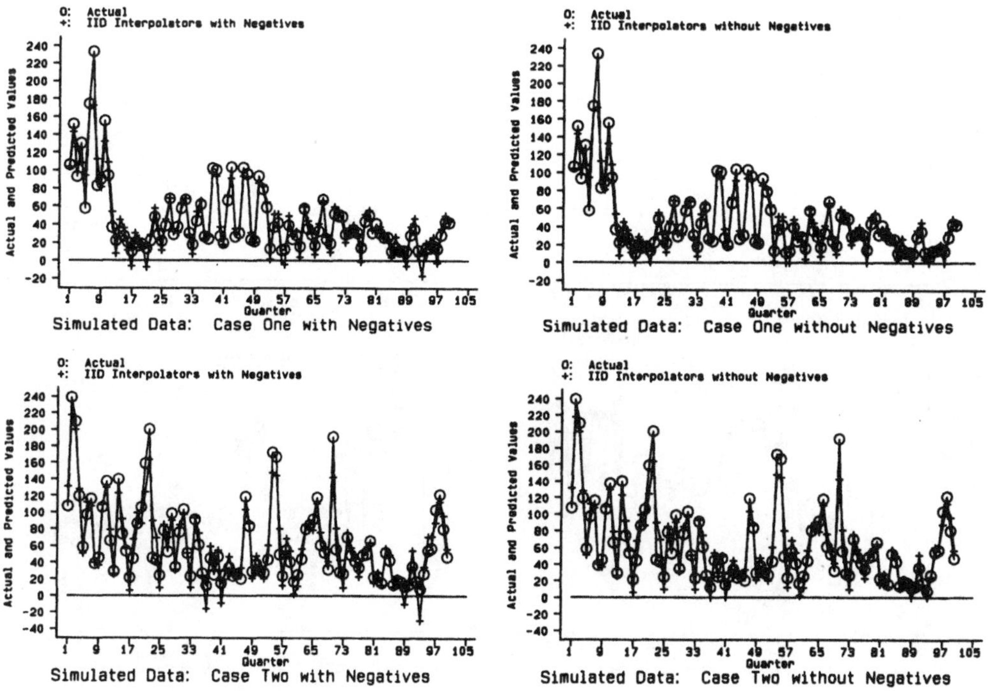

Figure 6.12:
German Iron & Steel Building Investment

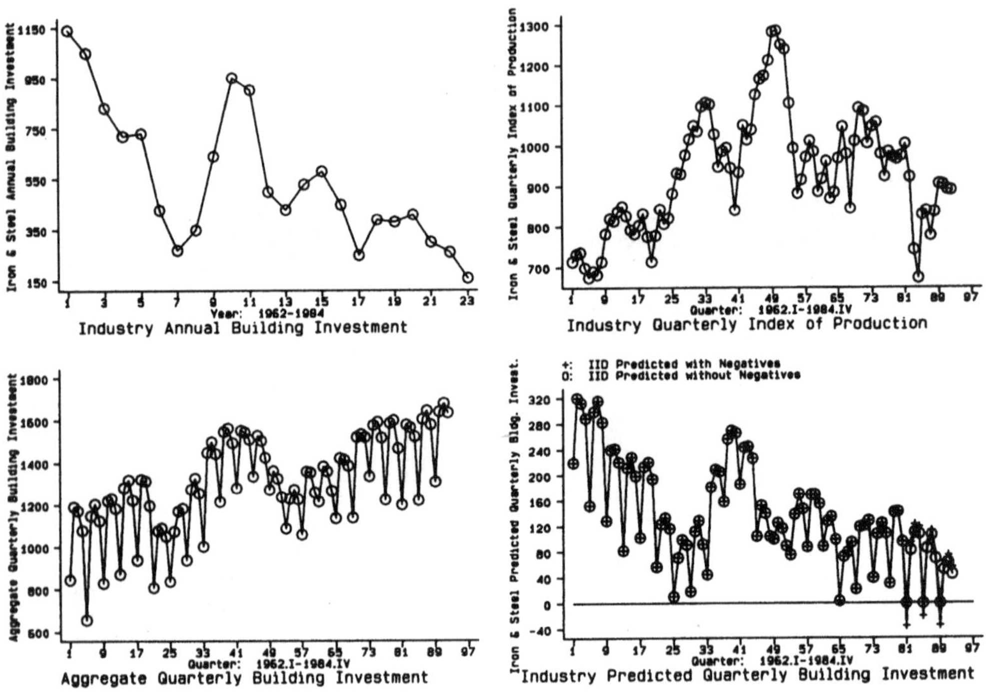

Figure 6.13:
German Rubber Products Equipment Investment

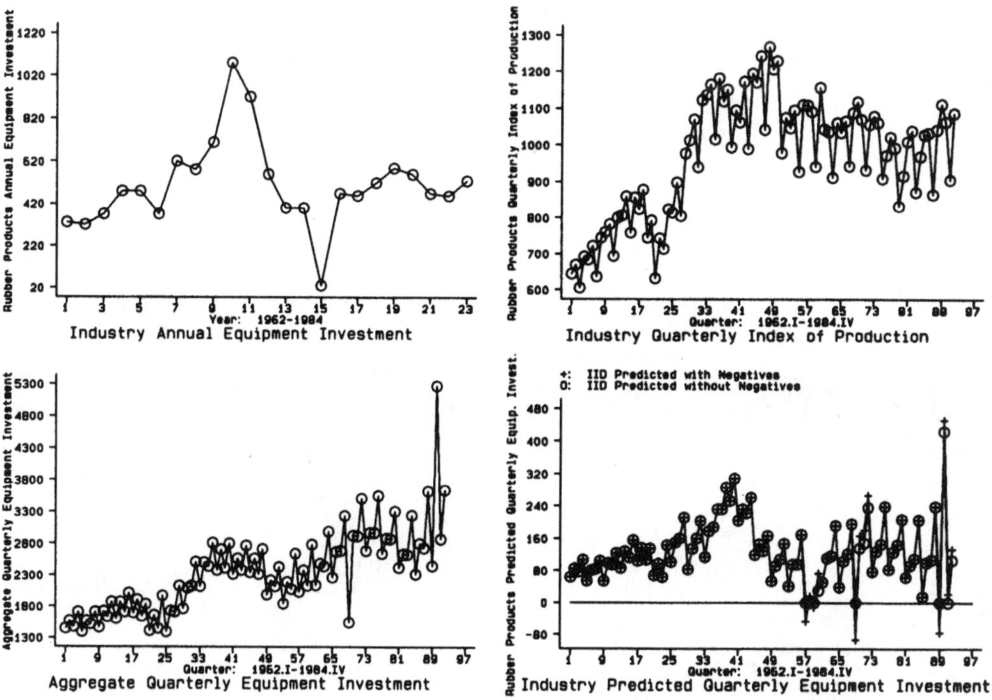

Figure 6.14:
Swedish Transportation Building Investment

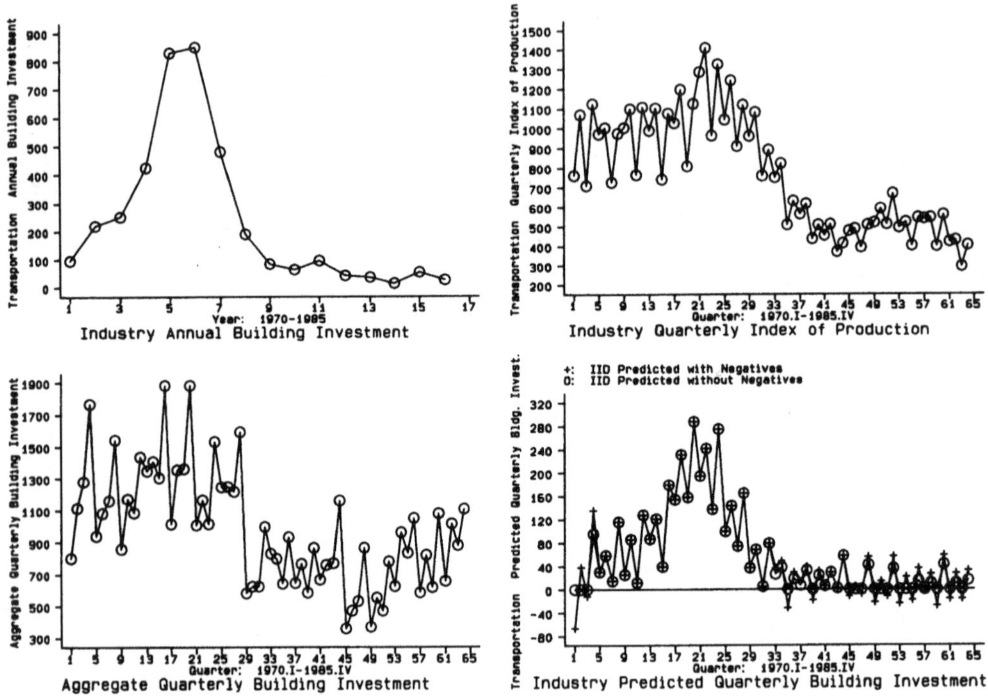

Figure 6.15:
Swedish Petroleum Products Equipment Investment

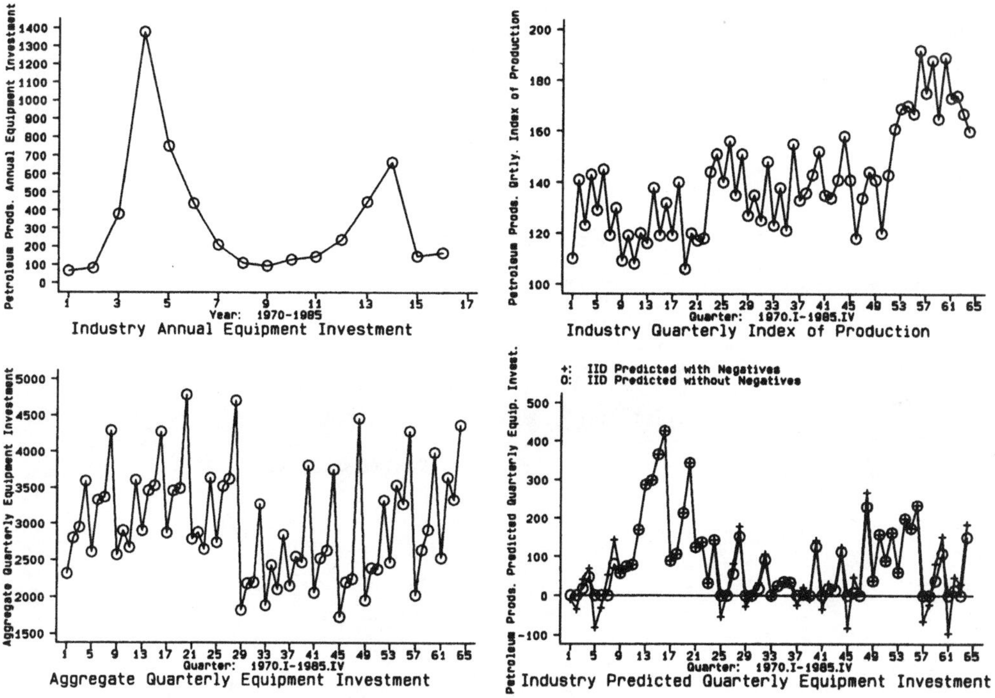

Figure 6.16:
German Import Price Indices

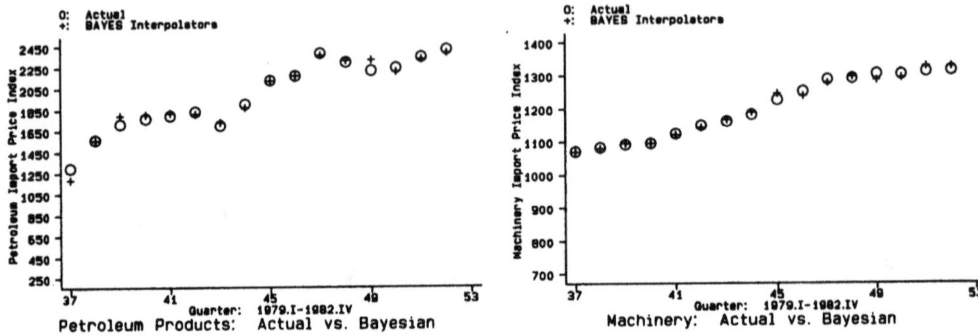

Figure 6.17:
German Export Price Indices

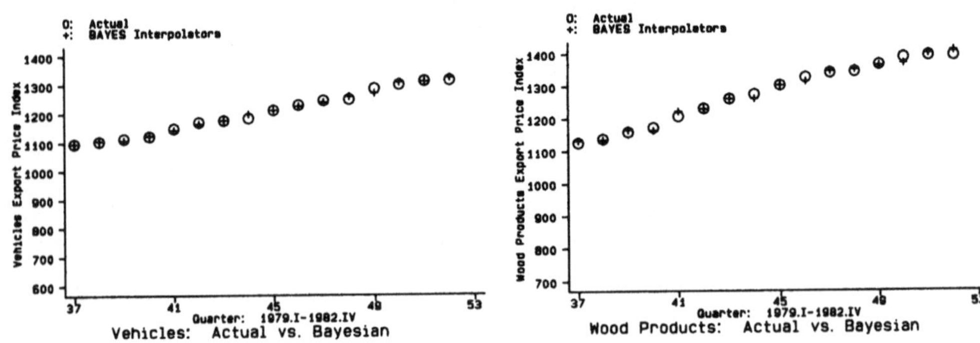

Figure 6.18:
German Hours Worked

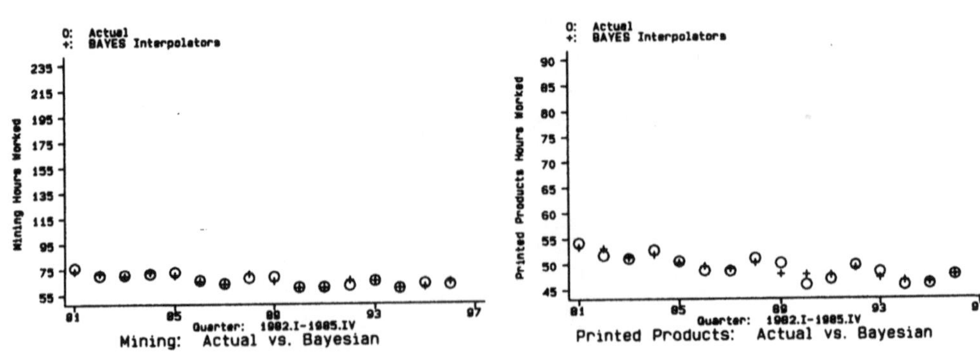

Figure 6.19:
German Employment

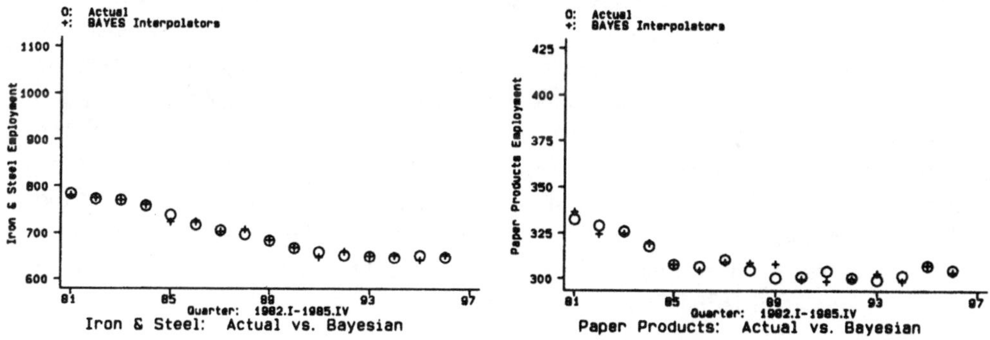

Figure 6.20:
German Sales Volumes

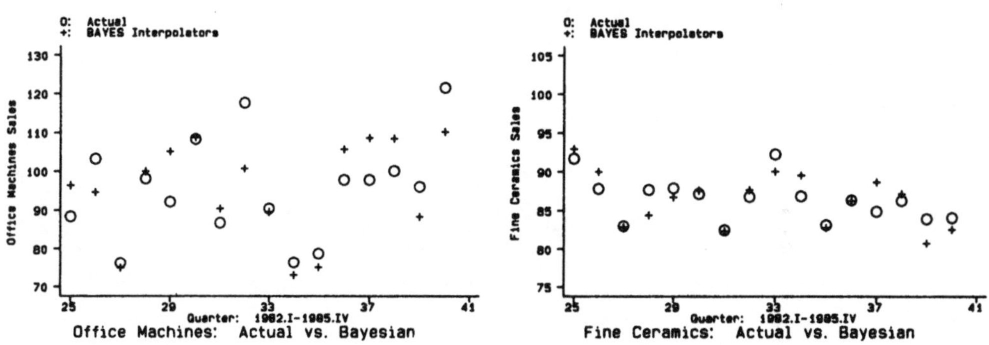

4 Conclusion

We have found that the standard Chow-Lin procedure provides good interpolations for a number of German data sets covering reasonably long time spans, and appears to be largely free of the so-called discontinuity problem according to which jumps may be expected to be observed from one low-frequency period to the next. Its main disadvantage, that it permits negative interpolated values, has been surmounted by implementation of a quadratic-programming algorithm which need only be applied separately to each low-frequency period for which negative interpolated values are obtained. This procedure is found to provide good interpolations in a simulated data set, and has been applied to German and Swedish investment data.

Finally, for short periods, a Bayesian procedure using prior information from an earlier data set performs better than standard procedures using information from the entire data set. This suggests that for countries that have only short and low-frequency time series, use of prior information and possibly also high-frequency related series from other countries could be expected to provide fairly reliable interpolated high-frequency series.

APPENDIX A: Basic Analytic Results and Proofs

This appendix collects the main analytic results used in the paper.

We start with a general Bayesian formulation of the interpolation problem. Given that the underlying high-frequency series are assumed to follow the relation (6.2) and are related to the low-frequency series by (6.3), we seek an affine function

$$\hat{y} = Az + c \tag{6.A1}$$

that best approximates y in terms of the criterion of mean-square error, i.e., that solves the problem

$$\min_{A,c} E(\hat{y} - y)'(\hat{y} - y) \tag{6.A2}$$

subject to (6.2) and (6.3). For this problem to be well-posed we must assume that β is a random variable with given mean and variance. Accordingly, we assume that β and ε have a joint prior distribution with mean and variance given by

$$E\begin{bmatrix}\beta\\ \varepsilon\end{bmatrix} = \begin{bmatrix}\bar{\beta}\\ 0\end{bmatrix} \quad \text{and} \quad \mathrm{Var}\begin{bmatrix}\beta\\ \varepsilon\end{bmatrix} = \begin{bmatrix}U & 0\\ 0 & V\end{bmatrix}. \tag{6.A3}$$

The optimal $\hat{y}$ is then termed the *best linear (affine) interpolator* of y, or equivalently, the wide-sense conditional expectation of y given z, denoted $\hat{E}(y|z)$ (cf. Doob [1953], pp. 77, 155). If β and ε are jointly normally distributed, this reduces to the ordinary conditional expectation, which is the posterior mean of y.

Theorem 1. The $T \times S$ matrix A and $T \times 1$ vector c which minimize

$$E(y - Az - c)'(y - Az - c) \tag{6.A4}$$

subject to

$$Gy = z \quad \text{and} \quad y = X\beta + \varepsilon, \tag{6.A5}$$

where β and ε have joint prior distribution with mean and variance given by (6.A3), are given by

$$c = (I - AG)X\bar{\beta} \tag{6.A6}$$

and

$$A = (XUX' + V)G'[G(XUX' + V)G']^{-1} \tag{6.A7}$$

respectively. The latter may also be written in the form

$$\begin{aligned} A \;=\; & VG'(GVG')^{-1} + [I - VG'(GVG')^{-1}]X[X'G'(GVG')^{-1} + \\ & U^{-1}]^{-1}X'G'(GVG')^{-1}. \end{aligned} \tag{6.A8}$$

Proof. From the restrictions (6.A5) we may write

$$y - Az - c = (I - AG)[X(\beta - \bar{\beta}) + \varepsilon] + [(I - AG)X\bar{\beta} - c]$$

whence, using (6.A3), see immediately that

$$E(y - Az - c)'(y - Az - c) = \\ \operatorname{tr}\{(I-AG)(XUX'+V)(I - AG)'\}+[(I-AG)X\bar{\beta}-c]'[(I-AG)X\bar{\beta}-c].$$

Minimization of the second term with respect to c gives (6.A6). It remains to minimize the first term with respect to A.

Writing

$$Q = \operatorname{tr}\{(I - AG)(XUX' + V)(I - AG)'\} \tag{6.A9}$$

we have

$$dQ = -2 \operatorname{tr}\{dA \cdot G(XUX' + V)(I - AG)'\} = 0$$

for all dA if and only if

$$G(XUX' + V) = G(XUX' + V)G'A'.$$

Expression (6.A7) for A then follows.

To obtain expression (6.A8) we make use of the basic identity

$$X'G'(GVG')^{-1}G(XUX'+V)G' = [X'G'(GVG')^{-1}GX+U^{-1}]UX'G' \tag{6.A10}$$

(cf. Chipman [1964], p. 1106). Pre- and postmultiplying (6.A10) by

$[X'G'(GVG')^{-1}GX + U^{-1}]^{-1}$ and $[G(XUX' + V)G']^{-1}$ respectively, we obtain for the first additive term of (6.A7):

$$XUX'G'[G(XUX'+V)G']^{-1} = X[X'G'(GVG')^{-1}GX+U^{-1}]^{-1}X'G'(GVG')^{-1}. \tag{6.A11}$$

Now from (6.A10) we also have

$$UX'G' = [X'G'(GVG')^{-1}GX + U^{-1}]^{-1}X'G'(GVG')^{-1}G(XUX' + V)G'$$

whence

$$GXUX'G' = GX[X'G'(GVG')^{-1}GX+U^{-1}]^{-1}X'G'(GVG')^{-1}G(XUX'+V)G'$$

and thus

$$\begin{aligned} GVG' &= G(XUX' + V)G' - GXUX'G' \\ &= \{I - GX[X'G'(GVG')^{-1}GX + U^{-1}]^{-1}X'G'(GVG')^{-1}\} \\ &\quad G(XUX' + V)G'. \end{aligned} \tag{6.A12}$$

Pre- and postmultiplying both sides of (6.A12) by

$VG'(GVG')^{-1}$ and $[G(XUX'+V)G']^{-1}$ respectively, we obtain

$$\begin{aligned} &VG'[G(XUX'+V)G']^{-1} = \\ &\quad VG'(GVG')^{-1}\{I - GX[X'G'(GVG')^{-1}GX + U^{-1}]^{-1}X'G'(GVG')^{-1}\}. \end{aligned} \tag{6.A13}$$

Putting together (6.A7), (6.A11), and (6.A13) we obtain

$$A = X(GX)^{\oplus} + VG'(GVG')^{-1}[I - GX(GX)^{\oplus}], \tag{6.A14}$$

where

$$(GX)^{\oplus} = [X'G'(GVG')^{-1}GX + U^{-1}]^{-1}X'G'(GVG')^{-1} \tag{6.A15}$$

is the "optimal inverse" of GX relative to GVG' (cf. Foster, 1961). This is equivalent to (6.A8). □

The Bayes estimator of β in the model

$$z = GX\beta + G\varepsilon, \tag{6.A16}$$

where (6.A3) holds, is

$$\hat{\beta} = (GX)^{\oplus}z + [I - (GX)^{\oplus}GX]\bar{\beta} = \bar{\beta} + (GX)^{\oplus}(z - GX\bar{\beta}) \tag{6.A17}$$

(cf., e.g., Chipman [1964], p. 1107). Thus, the Bayes interpolator of y may be written

$$\hat{y} = VG'(GVG')^{-1}z + [I - VG'(GVG')^{-1}G]X\hat{\beta} = X\hat{\beta} + VG'(GVG')^{-1}(z - GX\hat{\beta}). \tag{6.A18}$$

This has exactly the same form as the Chow-Lin interpolator (6.6), except that the Bayes estimator (6.A17) replaces the BLU estimator (6.7).

Theorem 2. The $T \times S$ matrix A which solves the problem

$$\text{Minimize } \operatorname{tr}(I - AG)V(I - AG)' \quad \text{subject to } (I - AG)X = 0 \tag{6.A19}$$

is given by

$$\begin{aligned} A &= VG'(GVG')^{-1} + [I - VG'(GVG')^{-1}G]X[X'G'(GVG')^{-1}GX]^{-1} \\ &\quad X'G'(GVG')^{-1} \\ &= X(GX)^{\ddagger} + VG'(GVG')^{-1}[I - GX(GX)^{\ddagger}], \end{aligned} \tag{6.A20}$$

where

$$(GX)^{\ddagger} = [X'G'(GVG')^{-1}GX]^{-1}X'G'(GVG')^{-1} \tag{6.A21}$$

is the oblique generalized inverse of GX relative to GVG' (cf. Chipman, 1964).

Proof. Defining the Lagrangean expression

$$L = \text{tr}[(I - AG)V(I - AG)' + 2(I - AG)X\Lambda'], \tag{6.A22}$$

where Λ is a $T \times k$ matrix of Lagrangean multipliers, we have that

$$dL = -2 \text{ tr}\{dA[GV(I - AG)' + GX\Lambda']\} = 0$$

for all dA if and only if

$$G[V(I - AG)' + X\Lambda'] = G(V + X\Lambda') - GVG'A' = 0,$$

hence

$$A = (V + \Lambda X')G'(GVG')^{-1}. \tag{6.A23}$$

Substituting this in the equation $(I - AG)X = 0$ we obtain

$$\Lambda = [I - VG'(GVG')^{-1}G]X[X'G'(GVG')^{-1}GX]^{-1}. \tag{6.A24}$$

Substituting (6.A24) in (6.A23) we obtain finally (6.A20). □

We note from (6.A15) and (6.A21) that

$$\lim_{U\to\infty}(GX)^{\oplus} = \lim_{U^{-1}\to 0}(GX)^{\oplus} = (GX)^{\ddagger}$$

hence (6.A14) reduces to (6.A20) as the prior variance of β approaches infinity. Since what counts in (6.A15) is the *relative* magnitudes of $X'G'(GVG')^{-1}GX$ and U^{-1}, as the sample size becomes larger the first term will dominate the second and the Bayes interpolator will approach the BLU interpolator. For short time series the Bayes interpolator is therefore to be recommended.

We now take up nonnegativity constraints. The following results, the second of which was first observed by Andreas Hornstein, may be stated without proof.

Theorem 1*. The $T \times S$ matrix A which solves the problem of minimizing (6.A4) subject to (6.A5) and the inequality constraint $\hat{y} \geq 0$ is obtainable from the set of quadratic programming problems

$$\min_{\{a_t\}_{t=4s-3,\ldots,4s}} \sum_{t=4s-3}^{4s} [a_tG(XUX' + V)G'a_t' - 2a_tw_t] \tag{6.A25}$$

subject to

$$\begin{aligned} \sum_{t=4s-3}^{4s} a_t z &= z_s \\ a_t[z - GX\bar{\beta}] + x_t\bar{\beta} &\geq 0 \quad (t = 4s-3, \ldots, 4s) \end{aligned} \tag{6.A26}$$

for $s = 1, 2, \ldots, S$, where:

a_t = the tth row of A; x_t = the tth row of X;
w_t = the tth column of $G(XUX' + V)$.

Theorem 2*. The $T \times S$ matrix A which solves the problem of minimizing (6.A4) subject to (6.A5) and the unbiasedness constraint (C2), as well as the inequality constraint $\hat{y} \geq 0$, is obtainable from the set of quadratic programming problems

$$\min_{\{a_t\}_{t=4s-3,\ldots,4s}} \sum_{t=4s-3}^{4s} [a_t GVG'a_t' - 2a_t w_t] \tag{6.A27}$$

subject to

$$\begin{aligned} a_t\bar{x}_k &= x_{tk} \quad (t = 4s-3, \ldots, 4s;\ k = 1, \ldots, K) \\ \sum_{t=4s-3}^{4s} a_t z &= z_s \\ a_t z &\geq 0 \quad (t = 4s-3, \ldots, 4s) \end{aligned} \tag{6.A28}$$

for $s = 1, 2, \ldots, S$, where:

a_t = the tth row of A; x_t = the tth row of X;
$\bar{x}_t$ = the tth row of GX; w_t = the tth column of GV.

APPENDIX B: Generation of Simulated Data

The method employed for generating high-frequency series which exhibit negatives in the interpolated values is described. Define the following notation:

$y =$ High-frequency series whose values are to be predicted using interpolation methods for $t = 1...100$.

$z_s =$ Low-frequency version of y for $s = 1, \ldots, 25$.

$x_{jt} =$ High-frequency related series $j = 1, 2$, for $t = 1, \ldots, 100$.

Generating these series proceeds in the following steps:

(1) Generate the related series x_{1t} and x_{2t} for $t = 1, \ldots, 100$.

(i) Using a random-number generator, generate u_{1t} and u_{2t} as follows:

$$\begin{bmatrix} u_{1t} \\ u_{2t} \end{bmatrix} = \begin{bmatrix} .5 & .3 \\ .3 & .5 \end{bmatrix} \begin{bmatrix} u_{1t-1} \\ u_{2t-1} \end{bmatrix} + \begin{bmatrix} \varepsilon_{1t} \\ \varepsilon_{2t} \end{bmatrix}$$

with $\varepsilon_{1t} \approx N(0, .0025)$ and $\varepsilon_{2t} \approx N(0, .01)$, iid and $u_{j1} = 0$ for $j = 1, 2$.

(ii) Using u_{1t} and u_{2t}, generate x_{1t} and x_{2t} as follows: $\forall t = 1, \ldots, 100$,

$$x_{1t} = \exp(u_{1t} + 8 - .5\log(t+3) - \mathcal{I}(t)) \qquad (6.B1)$$

$$x_{2t} = \exp(u_{2t} + 8 - .4\mathcal{I}(t)) \qquad (6.B2)$$

where

$$\mathcal{I}(t) = \begin{cases} .4 & \text{if } t = 1, 5, 9, \ldots \\ .2 & \text{if } t = 4, 8, 12, \ldots \\ 0 & \text{otherwise} \end{cases}$$

The $\log(t + 3)$ component in x_{1t} reflects downward trend in the series and the $\mathcal{I}(t)$ component in both x_{1t} and x_{2t} reflect seasonality. Note that x_{jt} is guaranteed to be positive for $j = 1, 2$ and $t = 1, \ldots, 100$.

(2) Generate the series of interest, y_t, as follows: $\forall t = 1, \ldots, 100$,

(i) Using a random-number generator and x_{jt} from above for $j = 1, 2$, generate w_t as follows: $\forall t = 1, \ldots, 100$,

$$w_t = \begin{bmatrix} x_{1t} \\ x_{2t} \end{bmatrix}' \begin{bmatrix} .001 \\ .0016 \end{bmatrix} - \log(t+3) + \eta_t$$

where $\eta_t \approx N(0, .0016)$, iid.

(ii) Generate y_t as follows: $\forall t = 1, \ldots, 100$,

$$y_t = \exp(w_t).$$

Note that y_t is guaranteed to be positive for $t = 1, \ldots, 100$.

(3) Generate the low-frequency series, z_s, from y_t as follows: $\forall s = 1, \ldots, 25$,

$$z_s = \sum_{t=4s-3}^{4s} y_t.$$

(4) Steps (1)–(3) are repeated twenty times to generate twenty sets of simulated data with properties consistent with the original investment data.

APPENDIX C: Data Sources

The following is a list of data sources for the series used in the paper. The German industry data follow the classification system *Güterverzeichniss für Produktionsstatistiken* (GP), Ausgabe 1982—a continuation of the former *Warenverzeichnis für die Industriestatistik* (WI). The Swedish series follow the classification system *Svensk standard för näringsgrensindelning* (SNI), second edition of the 1969 standard—a refinement of the United Nations *International Standard Industrial Classification of All Economic Activities* (ISIC). The frequency of the series is indicated in parentheses.

1. German import and export price indices (monthly), base 1976, 1970–1982. Statistisches Bundesamt, Wiesbaden. Fachserie 17, Reihe 8, *Preise und Preisindizes für die Ein- und Ausfuhr*, and unpublished data supplied by the Statistisches Bundesamt.

2. German import and export unit values (monthly), base 1976, 1970–1982. Statistisches Bundesamt, Wiesbaden. Unpublished data supplied to J.S.C. These series can also be obtained from published sources (but less accurately) by dividing import and export values by import and export volume indices respectively, and expressing the results relatively to the base year; the source for these is: Fachserie 7, Reihe 1, *Zusammenfassende Übersichten für den Außenhandel*, Spezialhandel, *Werte nach Warengruppen und -zweigen des Warenverzeichnisses für die Industriestatistik* and *Index des Volumens nach Warengruppen und -zweigen des Warenverzeichnisses für die Industriestatistik*.

3. German index of industrial production (monthly), base 1970, 1962–1976; base 1976, 1970–1979; base 1980, 1976–1985. Statistisches Bundesamt, Wiesbaden. Base 1970: Fachserie 4, Reihe 2.1, *Produktionsindizes, Index der Arbeitsproduktivität, Produktion ausgewählter industrieller Erzeugnisse. Indizes der industriellen Produktion. 1.1 Index der industriellen Nettoproduktion nach Industriegruppen und ausgewählten Zweigen, 1970=100. 1.1.2. von Kalenderunregelmäßigkeiten bereinigt.* Bases 1976 and 1980: Fachserie 4, Reihe 2.1, *Indizes der Produktion und der Arbeitsproduktivität, Produktion ausgewählter Erzeugnisse im Produzierenden Gewerbe. Index der Nettoproduktion für das produzierende Gewerbe, fachliche Unternehmensteile, von Kalenderunregelmäßigkeiten bereinigt.*

4. German monthly hours worked (monthly), 1970–1982. Statistisches Bundesamt, Wiesbaden. Fachserie 4, Reihe 4.1.1, *Beschäftigung, Umsatz und Energieversorgung der Unternehmen und Betriebe im Bergbau und im Verarbeitenden Gewerbe. Ergebnisse nach hauptbeteiligten Industriegruppen und -zweigen. 1.1. Betriebe, Beschäftigte, Arbeiterstunden, Lohn- und Gehaltssumme. Geleistete Arbeiterstunden (1000).*

5. German employment (monthly), 1970–1982. Statistisches Bundesamt, Wiesbaden. Fachserie 4, Reihe 4.1.1, *Beschäftigung, Umsatz und Energieversorgung der Unternehmen und Betriebe im Bergbau und im Verarbeitenden Gewerbe. Ergebnisse nach hauptbeteiligten Industriegruppen und -zweigen. 1.1. Betriebe, Beschäftigte, Arbeiterstunden, Lohn- und Gehaltssumme. Beschäftigte, insgesamt (Anzahl).*

6. German sales volumes in 1980 prices (monthly and quarterly), 1976–1985. Statistisches Bundesamt, Wiesbaden. Fachserie 4, Reihe 2.2, *Auftragseingang und Umsatz im Verarbeitenden Gewerbe, Auftragseingang und Auftragsbestand im Bauhauptgewerbe – Indizes –. Umsatz im Verarbeitenden Gewerbe, Volumenindex, 1980 = 100.*

7. German industry (annual) and aggregate (quarterly) building and equipment investment, 1962–1984. Deutsches Institut für Wirtschaftsforschung, Berlin, October 1989. *Sozialprodukt und Einkommenskreislauf I/1960 bis II/1989, Vierteljährliche volkswirtschaftliche Gesamtrechnung für die Bundesrepublik Deutschland. A. Entstehung, Verteilung und Verwendung des Bruttosozialprodukts. A3: Verwendung des Bruttosozialprodukts. A32: Zu Preisen von 1980.*

8. German building investment broken down by industry (annual), 1980 prices, 1960–1985. Statistisches Bundesamt, Wiesbaden, Fachserie 18, Reihe S.8, *Volkswirtschaftliche Gesamtrechnung, Revidierte Ergebnisse*

1960 bis 1984, table 3.18, *Anlage Investitionen nach Wirtschaftsbereichen, Neue Bauten, in Preisen von 1980.*

9. German equipment investment broken down by industry (annual), 1980 prices, 1960–1985. Statistisches Bundesamt, Wiesbaden. Fachserie 18, Reihe S.8, *Volkswirtschaftliche Gesamtrechnung, Revidierte Ergebnisse 1960 bis 1984*, table 3.18, *Anlage Investitionen nach Wirtschaftsbereichen, Neue Ausrüstungen, in Preisen von 1980.*

10. Swedish industry (annual) and aggregate (quarterly) building investment, 1970–1985. Statistiska centralbyrån, Stockholm. Statistiska Meddelanden, Table 1B: *Bruttoinvesteringar efter kapitaltyp och näringsren* (Gross fixed capital formation by type of asset and kind of economic activity), Table A: *Byggnader och anläggnigar* (Building and construction), millions of kronor, 1980 prices.

11. Swedish industry (annual) and aggregate (quarterly) equipment investment, 1970–1985. Statistiska centralbyrån, Stockholm. Statistiska Meddelanden, Table 1B: *Bruttoinvesteringar efter kapitaltyp och näringsren* (Gross fixed capital formation by type of asset and kind of economic activity), Table B: *Maskiner, transportmedel, inventarier* (Machinery, means of transport, equipment), millions of kronor, 1980 prices.

12. Swedish index of industrial production (monthly), 1970–1985, base 1980. Statistiska centralbyrån, Stockholm. Magnetic tape supplied to J.S.C.

References

Akhiezer, N. I. (1962): "The Calculus of Variations." New York: Blaisdell Publishing Company.

Chipman, John S. (1964): "On Least Squares with Insufficient Observations," *Journal of the American Statistical Association*, **59** (December): 1078–1111.

Chow, Gregory C. and An-loh Lin (1971): "Best Linear Unbiased Interpolation, Distribution, and Extrapolation of Time Series by Related Series," *Review of Economics and Statistics*, **53** (November): 372–375.

Doob, J. L. (1953): "Stochastic Processes." New York: Wiley.

Fernández, Roque B. (1981): "A Methodological Note on the Estimation of Time Series," *Review of Economics and Statistics*, **63** (August): 471–476.

Fisher, Sir Ronald A. (1958): "Statistical Methods for Research Workers." 13th edition. New York: Hafner Publishing Company, Inc.

Foster, Manus (1961): "An Application of the Wiener-Kolmogorov Smoothing Theory to Matrix Inversion," *Journal of the Society for Industrial and Applied Mathematics*, **9**: 387–392.

Friedman, Milton (1962): "The Interpolation of Time Series by Related Series," *Journal of the American Statistical Association*, **57** (December): 729–757.

Goldfarb, D. and A. Idnani (1983): "A Numerically Stable Dual Method for Solving Strictly Convex Quadratic Programs." *Mathematical Programming*, **27**: 1–33.

Haitovsky, Yoel (1973): "Regression Estimation from Grouped Observations." London: Griffin.

Harvey, A. C. and R. G. Pierse (1984): "Estimating Missing Observations in Economic Time-Series Analysis." *Journal of the American Statistical Association*, **79** (March): 125–131.

Hillmer, Steven C. and Abdelwahed Trabelsi (1987): "Benchmarking of Economic Time Series." *Journal of the American Statistical Association*, **82** (December): 1064–1071.

Kravis, Irving and Robert E. Lipsey (1974): "International Trade Prices and Price Proxies." In: Nancy D. Ruggles (ed.), *The Role of the Computer in Economic and Social Research in Latin America*. New York: National Bureau of Economic Research, pp. 253–268.

Kruskal, William (1968): "When are Gauss-Markov and Least Squares Estimators Identical? A Coordinate-Free Approach." *Annals of Mathematical Statistics*, **39** (February): 70–75.

Litterman, Robert B. (1981): "A Random Walk, Markov Model for the Interpolation of Time Series." Federal Reserve Bank of Minneapolis, Research Department Working Paper, December.

Powell, M. J. D. (1983a): "On the Quadratic Programming Algorithm of Goldfarb and Idnani." DAMTP Report NA19, Cambridge, England.

Powell, M. J. D. (1983b): "ZQPCVX: A FORTRAN Subroutine for Convex Quadratic Programming," DAMTP Report NA17, Cambridge, England.

Powell, M. J. D. (1989): "TOLMIN: A Fortran Package for Linearly Constrained Optimization Calculations," DAMTP Report 1989/NA2, University of Cambridge.

Prais, S. J. and J. Aitchison (1953): "The Grouping of Observations in Regression Analysis." *Revue de l'Institut International de Statistique*, **22**: 1–22.

Whittaker, Edmund and G. Robinson (1924): "The Calculus of Observations." London and Glasgow: Blackie & Sons Limited.

Chapter 7

Trend Interpolation and the Persistence of Fluctuations in U.S. GNP*

ALBERT JAEGER AND ROBERT M. KUNST

1 Introduction

Two recent papers by Cochrane [1988] and Cogley [1990] examined the persistence of fluctuations in real GNP. Both papers find little persistence in real U.S. per capita GNP series covering the periods 1869-1986 (Cochrane) and 1871-1985 (Cogley). Campbell and Mankiw [1987], however, argue that the persistence of U.S. post World War II GNP is substantial. Furthermore, comparing the persistence of real output series for various countries in the prewar and postwar era, DeLong and Summers [1988] report that first, prewar output series are in general less persistent than postwar series, and second, U.S. prewar GNP exhibits much lower persistence than comparable prewar output series for Sweden and the United Kingdom. These puzzling findings raise a natural question: Why is the persistence of U.S. prewar GNP fluctuations so unusually low?

Estimates of prewar U.S. GNP had to be derived from a sketchy data base with reliable data available only for benchmark years. The GNP values for in-between years were constructed by linear interpolation. In this note we examine the hypothesis that trend interpolation accounts for the reported low

*We would like to thank our discussant John Chipman for helpful remarks on an earlier version of this note.

persistence of U.S. prewar GNP. Stock and Watson [1986] conjectured that trend interpolation may bias the outcome of unit root tests. Cochrane [1988] briefly discussed the possible bias in persistence estimates due to measurement errors in prewar output data. In a closely related paper, Jaeger [1990] reported Monte Carlo simulations to illustrate the effects of trend interpolation on persistence estimates.

The note proceeds in three steps: First, we describe the trend interpolation model. Second, we explore analytically the effects of trend interpolation on the persistence of a GNP process which follows a random walk. And third, we report the results of an interpolation experiment based on U.S. postwar data.

2 The Trend Interpolation Model

The use of trend interpolation is well exemplified in recent papers providing new estimates of U.S. prewar GNP by Balke and Gordon [1989] and Romer [1989]. The main goal of both papers is to correct for excess cyclical volatility in U.S. prewar output series constructed by Kuznets [1961]. Both papers base their new estimates of U.S. prewar GNP on the assumption that the long run trend in GNP is captured by a piecewise linear trend which is interpolated between benchmark years.

To evaluate the effects of trend interpolation on persistence measures we consider the interpolation approach applied in Romer [1989]. Assume there are two time series y_t and x_t. y_t stands for the logarithm of real GNP, and the series is assumed to be unobservable except for certain benchmark years. x_t denotes the logarithm of real commodity output and is assumed to be observable in each time period. An estimate of GNP (y_t^*) is constructed by forming

$$y_t^* = \overline{y}_t + \beta(x_t - \overline{x}_t), \tag{7.1}$$

where $\overline{y}_t$ and $\overline{x}_t$ represent GNP and commodity output series, respectively, derived by linearly interpolating between benchmark years, and β is a constant interpolation parameter. The purpose of this note is to answer the question: How persistent are fluctuations in constructed GNP (y_t^*) compared to fluctuations in actual GNP (y_t) if actual GNP is in fact a persistent output series?

3 An Exploratory Example

We study an exploratory example to gauge the effects on persistence estimates when GNP data are constructed using equation (7.1). Assume the actual GNP and commodity output series are generated by the drift-adjusted random walk processes

$$y_t = y_{t-1} + u_t, \quad (7.2)$$
$$x_t = x_{t-1} + v_t. \quad (7.3)$$

The variances of the innovations u_t and v_t are σ_u^2 and σ_v^2 respectively. The contemporaneous covariance between the innovations may be nonzero and is denoted by σ_{uv}. We take the persistence measure of interest to be Cochrane's [1988] variance-ratio statistic

$$v = 1 + 2\sum_{j=1}^{\infty} r_j, \quad (7.4)$$

where r_j denotes the jth autocorrelation of the first difference of the logarithm of GNP. Thus, the persistence of actual GNP generated by equation (7.2) is 1.0. We further assume that the actual GNP series is observed only every n periods. The trend series $(\overline{v}_t)$ is constructed by interpolating linearly between the observed benchmark GNP values

$$\overline{y}_t = y_{d(t)} + (1/n)(t - d(t))(y_{d(t)+n} - y_{d(t)}), \quad (7.5)$$

where $d(t) = n[t/n]$, $t = 0, 1, \cdots$, and $[.]$ is the entier function. A trend series for commodity output $\overline{x}_t$ is calculated by using an analogous equation. Given the trend series for GNP and commodity output and a value for the interpolation parameter β, equation (7.1) is employed to construct an estimate of GNP.

To calculate the variance-ratio statistic for the constructed GNP series, we need the autocovariances of the first difference of y_t^* (the calculations are contained in the appendix)

$$c_j = \begin{cases} n^{-1}\sigma_u^2 + \beta^2((n-1)/n)\sigma_v^2 & \text{for} \quad j = 0, \\ ((n-j)/n^2)(\sigma_u^2 - \beta^2\sigma_v^2) & \text{for} \quad 1 \le j < n \\ 0 & \text{for} \quad j \ge n. \end{cases} \quad (7.6)$$

Inserting the autocorrelations based on (7.6) in equation (7.4), the persistence of the constructed GNP series is

$$V = \frac{n\sigma_u^2}{\sigma_u^2 + \beta^2(n-1)\sigma_v^2}. \quad (7.7)$$

For reasonable values of n, a well-working approximation for equation (7.7) is $V \approx \sigma_u^2/\beta^2\sigma_v^2$. We note two interesting cases: First, if $\beta = 1.0$, the persistence of constructed GNP will approximate σ_u^2/σ_v^2. If the volatility of commodity output (σ_v^2) exceeds the volatility of the actual GNP series (σ_u^2), the persistence of constructed GNP will be smaller than 1.0. Second, if $\beta = \sigma_u/\sigma_v$, the persistence of actual and reconstructed GNP will coincide. Thus, the choice of the interpolation parameter β is crucial for the extent of bias in persistence estimates introduced by interpolation. We note from the expression for the variance of the growth rate of constructed GNP in equation (7.6) that setting $\beta = 1.0$ will overstate volatility if commodity output fluctuates more than GNP and that setting $\beta = \sigma_u/\sigma_v$ reproduces the correct volatility of actual GNP growth rates.

4 An Interpolation Experiment

In this section we report the results of an interpolation experiment based on U.S. postwar data to illustrate the practical usefulness of our exploratory example. We assume hypothetically that real annual U.S. postwar GNP is only observed for the years 1947, 1955, 1962, 1972, 1981, and 1989, whereas a real commodity output series is observed annually over the full sample period 1947–1989. The choice of benchmark years follows Romer [1989]. The data source is the *Economic Report of the President* [1990]. Real annual GNP is from Table C-2 and commodity output is approximated by total industrial production taken from Table C-49.

Equation (7.1) was used to construct an estimate of postwar U.S. GNP for a range of values of the interpolation parameter β. Table 7.1 compares the volatility and persistence of the actual and the constructed GNP series. The volatility is measured by the standard deviation of growth rates in percent, and persistence is measured by the small sample variance-ratio statistic suggested by Cochrane [1988] as an approximation for the persistence measure defined in equation (7.4).

For our purposes, the most interesting case in table 7.1 is the benchmark case $\beta = 1.0$. Compared to the results for the actual GNP series, persistence of the constructed GNP series is about 50 percent lower. Cochrane [1988] took the pre-1939 output data from Friedman and Schwartz [1982], which are in turn mainly based on Kuznets's work (e.g. Kuznets [1961]). Romer [1989] argues that the Kuznets data overstate the volatility of prewar business cycles because Kuznets implicitly assumed b=1.0. Thus, the low persistence estimates reported by Cochrane [1988] might be consistent with substantial persistence in actual U.S. prewar GNP. More generally, if data sources for constructing prewar output series were fragmentary in other countries and Kuznets-type interpolation was used to construct prewar output data, one would expect to find low persistence (high volatility) estimates for these countries. On the other hand, output series based on better data sources for the prewar period and/or constructed with less reliance on Kuznets-type interpolation than in the U.S. case, should exhibit larger persistence (and lower volatility) than U.S. data. This observation may explain why prewar Swedish and U.K. output data are more persistent than U.S. prewar output data as reported by DeLong and Summers [1988].

Table 7.1:

Persistence and Interpolation:
An Experiment Based on U.S. Postwar Data

GNP Series	Stdv. in %	Variance-Ratio k=4	k=8	Statistic k=16
Actual GNP Series	2.59	1.06	0.81	0.81
Constructed GNP Series				
β=1.00	5.44	0.55	0.43	0.43
β=0.90	4.91	0.56	0.45	0.45
β=0.80	4.37	0.58	0.47	0.48
β=0.70	3.84	0.60	0.50	0.53
β=0.60	3.31	0.64	0.54	0.59
β=0.50	2.78	0.69	0.61	0.68
β=0.40	2.26	0.78	0.73	0.82
β=0.30	1.75	0.95	0.96	1.07
β=0.20	1.27	1.33	1.46	1.57
β=0.10	0.86	2.22	2.66	2.60

Notes: The small sample variance ratio statistic is calculated as $(k^{-1}var(y_t - y_{t-k})/var(y_t - y_{t-1}))(T/(T-k+1))$, where y_t is the logarithm of GNP, and T is the sample size. β is the interpolation parameter.

If the interpolation parameter β is reduced below 1.0 in table 7.1, the persistence of the constructed GNP series increases nonlinearly. Persistence estimates for constructed GNP series based on the assumption that $\beta < 1.0$ may therefore depend sensitively on the exact value chosen for the interpolation parameter.

5 Conclusions

The note argued that widely reported persistence and volatility properties for U.S. prewar GNP data may be due to the use of trend interpolation techniques in the data construction process. We suspect that trend interpolation is more widely used than suggested by the prewar data example analyzed in this note. Statistical bureaus appear to use trend interpolation quite routinely for constructing data series where primary data sources are sketchy or the availability of good data is restricted to benchmark values.

APPENDIX

Derivation of the autocovariances of constructed GNP

Start with the first difference of the logarithm of constructed GNP

$$\Delta y_t^* = \Delta \overline{y}_t + \beta(\Delta x_t - \Delta \overline{x}_t) \tag{7.A1}$$

If y_t and x_t are generated by random walk processes driven by the innovations u_t and v_t, respectively, the interpolation formula (7.5) in the main text implies

$$\Delta \overline{y}_t = n^{-1}(u_{d(t-1)+1} + \cdots + u_{d(t-1)+n}), \tag{7.A2}$$

$$(\Delta x_t - \Delta \overline{x}_t) = v_t - n^{-1}(v_{d(t-1)+1} + \cdots v_{d(t-1)+n}). \tag{7.A3}$$

where $d(t) = n[t/n]$, $t = 1, \ldots,$ and $[.]$ is the entier function. The autocovariances of Δy_t^* depend on the second moments of $\Delta \overline{y}_t$ and $(\Delta x_t - \Delta \overline{x}_t)$. The variance of $\Delta \overline{y}_t$ is

$$E[(\Delta \overline{y}_t)^2] = n^{-1}\sigma_u^2. \tag{7.A4}$$

The first-order autocovariance of $\Delta \overline{y}_t$ is given as

$$E[\Delta \overline{y}_t \Delta \overline{y}_{t-1}] = \begin{cases} n^{-1}\sigma^2 & \text{for } t \neq d(t)+1, \\ 0 & \text{for } t = d(t)+1. \end{cases} \tag{7.A5}$$

The interpolated time series $\overline{y}_t$ is therefore not stationary after first differencing. We define the large-sample autocovariance of a mean-zero random variable z_t with time-dependent autocovariances as

$$c(z_t, z_{t-1}) = \lim_{T\to\infty} T^{-1} \sum_{t=j+1}^{T} z_t z_{t-j}, \tag{7.A6}$$

where T is the sample size. Using the concept of large-sample autocovariances defined by (7.A6), we may average over the time-dependent autocovariances in (7.A5) using the weights $n-1$ and 1

$$c(\Delta\overline{y}_t, \Delta\overline{y}_{t-1}) = ((n-1)/n^2)\sigma_u^2, \tag{7.A7}$$

Analogous calculations for $2 \leq j < n$ result in

$$c(\Delta\overline{y}_t, \Delta\overline{y}_{t-1}) = ((n-j)/n^2)\sigma_u^2, \tag{7.A8}$$

and the autocovariances for $j \geq n$ are zero. The variance of $(\Delta x_t - \Delta\overline{x}_t)$ is

$$E[(\Delta x_t - \Delta\overline{x}_t)^2] = ((n-1)/n)\sigma_v^2, \tag{7.A9}$$

For the autocovariances of $\Delta x_t - \Delta\overline{x}_t$ for $l \leq j < n$, the same procedure used for the calculation of the autocovariances of $\Delta\overline{y}_t$ yields

$$c((\Delta x_t - \Delta\overline{x}_t), (\Delta x_{t-j} - \Delta\overline{x}_{t-j})) = \begin{cases} -((n-j)/n^2)\sigma_v^2 & \text{for } 1 \leq j < n, \\ 0 & \text{for } j \geq n. \end{cases} \tag{7.A10}$$

Finally, the crosscovariances between $\Delta\overline{y}_t$ and $(\Delta x_t - \Delta\overline{x}_t)$ are all zero. Combining (7.A4), (7.A7), (7.A8), (7.A9), and (7.A10), gives equation (7.6) in the main text.

References

Balke, N. S., R. J. Gordon (1989): "The estimation of prewar gross national product: Methodology and new evidence." *Journal of Political Economy*, **97**: 38–92.

Campbell, J. Y. and N. G. Mankiw (1987): "Are output fluctuations transitory?" *Quarterly Journal of Economics*, **102**: 857–880.

Cochrane, J. H. (1988): "How big is the random walk in GNP?." *Journal of Political Economy*, **96**: 893–920.

Cogley, T. (1990): "International evidence on the size of the random walk in output." *Journal of Political Economy*, **98**: 501–518.

DeLong B. J. and L. H. Summers (1988): "How does macroeconomic policy affect output?." *Brookings Papers on Economic Activity*, **2**: 433–480.

Friedman M. and A. J. Schwartz (1982): "Monetary Trends in the United States and the United Kingdom: Their Relation to Income, Prices, and Interest Rates 1867–1975." Chicago (University of Chicago Press).

Jaeger, A. (1990): "Shock persistence and the measurement of prewar output series." *Economics Letters*, **43**: 33–37.

Kuznets, S. S. (1961): "Capital in the American Economy: Its Formation and Financing." Princeton (Princeton University Press).

Romer C. D. (1989): "The prewar business Cycle reconsidered: New estimates of gross national product, 1869–1908." *Journal of Political Economy*, **97**: 1–37.

Stock J. H. and M. W. Watson (1986): "Does GNP have a unit root?." *Economics Letters*, **22**: 147–151.

Chapter 8

Short-Term Forecasts of the Basic Economic Indicators for the Polish Economy

NINA ŁAPIŃSKA–SOBCZAK

1 Introduction

The necessity of short term analysis requires appropriate statistical data. The year became a period to long for such analysis. Quarterly and even monthly forecasts become necessary. Thus, a model satisfying targets of such forecasting can also be estimated on the basis of data for periods shorter than a year. Such models have existed for the countries with developed market economies for many years. The present macro models for the Polish economy and for other Eastern European countries have been based, first of all, on annual data. One of the main reasons of giving up the research was the lack of statistical data, hence, the inability to verify the model empirically. Since 1989 the statistical bulletins present monthly data, also for 1988, concerning the basic quantities of our economy. Hence, there is now an opportunity to estimate the model's parameters on the basis of monthly or quarterly series, (the transition to the market system, deep recession, high inflation rate). A model based on quarterly or monthly data should have revised specification as against the existing yearly model. Within such a model, more emphasis should also be put on its financial sector. The role of financial instruments in the market economy is very important, as they influence many real processes.

During the first stage of research, before a suitable model including causal relations, is constructed, one should perform a statistical analysis of the obtained monthly time series. An appropriate data bank contains information from January of 1988. The analysis will be mostly concentrated on checking the behavior in time and, possibly, suggesting some simple methods of forecasting (time series extrapolation) these quantities for the future periods.

2 Characteristics of the Applied Forecasting Methods

In such cases much more correct is forecasting by using adaptive methods. They do not need tight assumptions on the stability of the analytical form and even the continuity of the relationship. They are also much more useful in cases when one needs to determine the forecasts concerning the shaping of variables characterized by a high level of irregularity and collapses in the trend. Hence, in these models one does not assume a constant analytical form of the relation between the endogenous variable y and the time variable t, but it is assumed that for each period the estimation of the trend level, and possibly seasonal fluctuations, is built as a mean of these estimations for the previous periods and values of the latest realizations of the endogenous variable. When determining these quantities higher weights are allocated to the "newer" information.

In the class of the adaptive models, the most common are the models of Brown exponential smoothing (Brown [1962]).

2.1 Forecasting of Processes Without Seasonality

Let us assume that for a given endogenous variable y we have the following n observations in time:

$$y_1, y_2, \ldots\ldots y_n.$$

If y is a time function then, generally, the model can be written:

$$y_t = f(t,\ \beta) + \varepsilon_t \tag{8.1}$$

assuming, that the random factor ε_t satisfies the classical assumptions. $f(t, \beta)$ can take different functional forms. In the simplest case, it is assumed that the form is as follows:

$$y_t = \beta + \varepsilon_t \tag{8.2}$$

This is the so-called constant mean model. If the parameter β is known (estimated), then the forecast for period $n + h$ is given

$$\hat{y}_n(h) = \hat{\beta} \tag{8.3}$$

The value of β is estimated as the arithmetical mean from the sample:

$$\hat{\beta} = \overline{y} = \frac{1}{n}\sum y_t \tag{8.4}$$

then value of the forecast is:

$$\hat{y}_n(h) = \overline{y} \tag{8.5}$$

hence, it is uniform for all periods of the forecast. This model can be used in forecasting processes, where particular observations are independent from the mean value, i.e. the coefficient of the residuals auto correlation r_a should not excess a doubled standard error (Abraham and Ledolter [1983]).

To examine whether forecasts are in line with the actual realization of a given process, the above shown rule of forecasting can be also applied within the sample period. For calculating the arithmetic mean at least two observations ($n = 2$) are necessary, therefore, the first forecast $\hat{y}_{n+1}$ can be determined for the third period. Generally the formula is written as follows:

$$\begin{aligned} \hat{y}_{n+1}(h) &= \frac{1}{(n+1)}(y_1 + y_2 + \cdots + y_n + y_{n+1}) \\ &= \frac{n}{n+1}\hat{y}_n(h) + \frac{1}{n+1}y_{n+1} \\ &= \hat{y}_n(h) + \frac{1}{n+1}\left[y_{n+1} - \hat{y}_n(h)\right]. \end{aligned} \tag{8.6}$$

It results from the above that a forecast of the variable for period $n+1$ is equal to the forecast for the previous period plus an adjustment term associated with the difference between the actual realization of y and its forecast from the previous period. The stated forecasting formula assumes, that all observations within the process are taken with equal weights. Very often, however, it seems more reasonable to use different weights in such a way that the latest observations are given higher weights while older observations get lower weights.

A notion of a so-called smoothing statistics or smoothing values S is being introduced here:

$$S_n = S_n^1 = (1-\omega)\left[y_n + \omega y_{n-1} + \omega^2 y_{n-2} + \cdots\right]. \tag{8.7}$$

This statistics is a forecast for the future periods, i.e.:

$$\hat{y}_n(h) = S_n$$

In case of a so-called simple smoothing:

$$S_n = \alpha y_n + (1-\alpha)S_{n-1} \tag{8.8}$$

$$\hat{y}_n(h) = \alpha y_n + (1-\alpha)S_{n-1} \tag{8.9}$$

where: $\alpha = 1 - \omega$. The initial value of the smoothing statistics S_1 is the value of y_1. Parameter β is defined as the smoothing parameter and, precisely, it is the weight allocated to the last observation. It takes values from the interval (0,1), depending on whether we want to allocate a higher weight to the latest realization of variable y or all the previous information is to be of greater significance. With a high level of α, forecasts react fast to changes in the last observations.

In practical applications the size of a parameter is determined empirically, assuming level which ensures, for example, the least ex post errors of a forecast. Most often, the parameter takes values from the interval 0.70 - 0.95.

Errors considered in the estimation of the a level are as follows:

ME	–	mean error
MSE	–	mean square error
MAE	–	mean absolute error
MPE	–	mean percentage error
MAPE	–	mean absolute percentage error

In the second approach it is assumed that variable y_t is shaped according to the following relation:

$$y_t = \beta_0 + \beta_1 t + \varepsilon_t \tag{8.10}$$

Parameters β_i are in most cases unknown. They can be estimated using, for example, the least squares method on the basis of past observations. Forecasts of variable y can be written then as:

$$\hat{y}_n(h) = \beta_{0,n} + \beta_{1,n} h \tag{8.11}$$

Introducing the adjustment statistics of order k:

$$S_n^k = (1-\omega)S_n^{(k-1)} + \omega S_{n-1}^k \tag{8.12}$$

which means that the smoothing process is applied k times. Taking advantage of the above definition of statistics S_n one can also derive formulas for parameters estimates:

$$\beta_{0,n} = 2S_n^1 - S_n^2; \qquad \beta_{1,n} = \frac{1-\omega}{\omega}\left[S_n^1 - S_n^2\right]. \tag{8.13}$$

Substituting them to the formulas which describe forecasts we have:

$$\hat{y}_n(h) = \left(2 + \frac{\alpha}{1-\alpha}h\right) S_n^1 - \left(1 + \frac{\alpha}{1-\alpha}h\right) S_n^2 \tag{8.14}$$

This procedure is known as a *double* or *linear* exponential smoothing, which means that statistics S_n^1 is adjusted for a second period. Using the above forecasting procedure one does not actually need to estimate parameters β_i. In this case, the adjustment parameter α should be close to zero, although there are many deviations from this formula.

In the next approach we assume that the model is in the form of a square polynomial:

$$y_t = \beta_0 + \beta_1 t + \beta_2 \frac{t^2}{2} + \varepsilon_t \tag{8.15}$$

In this case the needed smoothing statistics are following:

$$\begin{aligned} S_n^1 &= (1-\omega)y_n + \omega S_{n-1}^1 \\ S_n^2 &= (1-\omega)S_n^1 + \omega S_{n-1}^2 \\ S_n^3 &= (1-\omega)S_n^2 + \omega S_{n-1}^3. \end{aligned}$$

So, the forecast is determined in accordance with the formula:

$$\begin{aligned} \hat{y}_n(h) &= \left[3 + \frac{\alpha(6-5\alpha)}{2\omega^2}h + \frac{\alpha^2}{2\omega^2}h^2\right] S_n^1 - \\ &- \left[3\frac{\alpha(5-4\alpha)}{\omega^2}h + \frac{\alpha^2}{\omega^2}h^2\right] S_n^2 + \\ &+ \left[1 + \frac{\alpha(4-3\alpha)}{2\omega^2}h + \frac{\alpha^2}{2\omega^2}h^2\right] S_n^3 \end{aligned} \tag{8.16}$$

and it is called a *triple* or *quadratic* exponential smoothing. The author of this method suggests choices of ω such that $0.7 < \omega^2 < 0.95$, which means that $0.89 < \omega < 0.98$.

2.2 Forecasting With Seasonality

Additive or multiplicative decomposition of the process is assumed:

$$y_t = T_t + S_t + \varepsilon_t \tag{8.17}$$

or

$$y_t = T_t * S_t * \varepsilon_t \tag{8.18}$$

where:

T_t – trend
S_t – seasonal index
e_t – random component

Before starting analyzing the process one must decide on the form of the preferred model. The seasonal component can be derived using some indicator, for example, dummy variables, indicator or some cyclic function, for example trigonometric functions: *sin, cos*. Therefore, a model with seasonality may have, for example, the following form:

$$y_t = \beta_0 + \sum_{i=1}^{k} \beta_i \frac{t_i}{i!} + \sum_{i=1}^{s} \delta_s \mathrm{IND}_{ti} + \varepsilon_i \tag{8.19}$$

or

$$y_t = \beta_0 + \sum_{i=1}^{k} \beta_i * \frac{t_i}{i!} + \sum_{i=1}^{m} A_i \sin\left(\frac{2\pi i}{S} + \Phi_i\right) + \varepsilon_i \tag{8.20}$$

In the first model there appear problems with parameters estimation due to the collinearity of explanatory variables. The problem can be defeated, however, in three ways:

- leaving out the free term β_0
- assuming that $\sum_{i=1}^{s} \delta_i = 0$
- assuming that one value of δ_s equals zero.

An extension of the exponential adjustment method to the case of seasonality is the Winters method. In the first approach it is assumed that model of the process is an additive one, in which the trend factor is linear:

$$T_{n+j} = \mu_{n+j} = \mu_n + \beta_j$$

and the random component adds up to 0. The process forecast is determined, in this case, according to the formula (Abraham and Ledolter [1983]):

$$\hat{y}_n(h) = \hat{\mu}_n + \hat{\beta}_n h + \hat{S}_{n+h-s} \tag{8.21}$$

Values $\hat{\mu}_n, \hat{\beta}_n$ and $\hat{S}_n$ are derived from the following relations:

$$\hat{\mu}_{n+1} = \alpha_1(y_{n+1} - \hat{S}_{n+1-S}) + (1-\alpha_1)(\hat{\mu}_n + \hat{\beta}_n) \tag{8.22}$$

$$\hat{\beta}_{n+1} = \alpha_2(\hat{\mu}_{n+1} - \hat{\mu}_n) + (1-\alpha_2)\hat{\beta}_n \tag{8.23}$$

$$\hat{S}_{n+1} = \alpha_3(\hat{y}_{n+1} - \hat{\mu}_{n+1}) + (1-\alpha_3)\hat{S}_{n+1-s} \tag{8.24}$$

The initial values for $\hat{\mu}$, $\hat{\beta}$ and $\hat{S}$ are most often estimated using least squares method, assuming that:

$$\hat{\mu}_0 = \hat{\mu}, \hat{\beta}_0 = \hat{\beta}, \text{ and } \hat{S}_0 = -\sum_{j=1}^{s-1} \delta_j$$

One can also consider the multiplicative Winters' model in the form:

$$y_{n+j} = (\mu_n + \beta t)S_{n+j} + \varepsilon_{n+j} \tag{8.25}$$

where: S is the multiplicative seasonal index, such that:

$$S_i = S_{i+s} = S_{i+2s} = \cdots \qquad \text{for} \quad i = 1, 2, \cdots s$$

and under the condition: $\sum_{i=1}^{s} S_i = s$. In this case the forecast is determined according to the formula:

$$\hat{y}_n(h) = (\hat{\mu}_n + \hat{\beta}_n h)\hat{S}_{n+h-s} \qquad \text{when} \quad h = 1, 2, \ldots s$$

or

$$\hat{y}_n(h) = (\hat{\mu}_n + \hat{\beta}_n)\hat{S}_{n+h-2s} \qquad \text{when} \quad h = s+1, \ldots 2s$$

etc. Elements of this relation are derived from relations similar to those which were given for the additive model, namely:

$$\hat{\mu}_{n+1} = \alpha_1 \frac{y_{n+1}}{\hat{S}_{n+1-s}} + (1-\alpha_1)(\hat{\mu}_n + \hat{\beta}_n)$$

$$\hat{\beta}_{n+1} = \alpha_2(\hat{\mu}_{n+1} - \hat{\mu}_n) + (1-\alpha_2)\hat{\beta}_n$$

$$\hat{S}_{n+1} = \alpha_3 \frac{y_{n+1}}{\mu_{n+1}} + (1-\alpha_3)\hat{S}_{n+1-s}$$

Smoothing parameters α_1, α_2, α_3 in both types of models take values from the interval (0,1). Their sizes are selected empirically so that the forecast errors were as low as possible. Generally, however, a growth in the values of one parameter should be accompanied by a fall in the others. Another case of the Winters' model is used in the STATGRAF and other packages, for example SAS. If we deprive the Winter's model of the seasonal index, then we obtain a forecasting method equivalent to Brown's *double* exponential smoothing. It is known in the literature as *Holt method.*

3 Empirical Analysis of Some Development Indicators of the Polish Economy

Statistical data characterizing the macroeconomic quantities of our economy in the monthly-quarterly arrangement are available only from 1988. Hence, we have now at our disposal (up to June 1991) a maximum of 30 observations. However, there are a lot of categories, for which data in the mentioned arrangement are not recorded, e.g. national income, total output, investments. There are also plenty of categories which are completely new to our economy (or some data were not recorded in the monthly arrangement) and information about them appeared much later, that is, only in 1989 or even 1990. This group includes most banking data.

Hence, the task of creating a set of information which would be used to estimate the parameters of a macro model of the economy in the transition period towards a market system is very tough.

The forecasting methods presented in the theoretical part of the paper will now be applied to examine the behavior of particular series in time. Certainly, this paper presents only some of them. Before presenting the results of forecasts, a preliminary analysis of the statistical data will examine a few indicators characterizing various sectors of our economy. At present, there will be eight such indicators discussed:

VYZP / VYP	–	households incomes from wages and salaries / total incomes of population [1],
VYZP / OLP	–	households incomes from wages and salaries / total savings of population [2],
VYZP / SQQP	–	households incomes from wages and salaries/ total sale of industrial products [3],
EP / MP	–	total exports /total imports [4],
ULBLA	–	unemployment rate [5],
BZLP/ OLP	–	banks indebtedness of households / bank deposits of households [6],
YCZP / YCP	–	expenditures on foods / total expenditures of households [7],
AFDP/ BCP	–	subsidies from state budget / total expenditures from state budget [8],

Values of these indicators are shown in figures 8.1–8.3 and in table 8.1. It is very easy to see that their values are not stable and they have a seasonal pattern. The period covering years from 1989 to the second quarter of 1991 was characterized by non homogeneous economic policy. At the second half of the year 1989 began the process of transferring into a market economy, but its mechanisms could not be observed earlier than in 1990. At the same time financial instruments started functioning. High inflation accompanied the process of transformation into market economy. This phenomenon has not been stopped until the end of 1990. Figure 8.1 presents household incomes from wages and salaries relating to the other variables: total income of these household, total bank deposits of households and the value of sales of industrial products. The last indicator presents the highest variability mainly due to the strong sales decrease (at the end of 1989 and the beginning of 1990). In the second half of 1990 a systematic decrease of income from wages and salaries occurs, which can be explained by privatization of Polish firms. The sudden increase of this indicator during first months of 1991 has been caused by extra incomes from the profit concerning the previous year.

Figure 8.1:
Shares of Income of Households in Other Economic Indicators

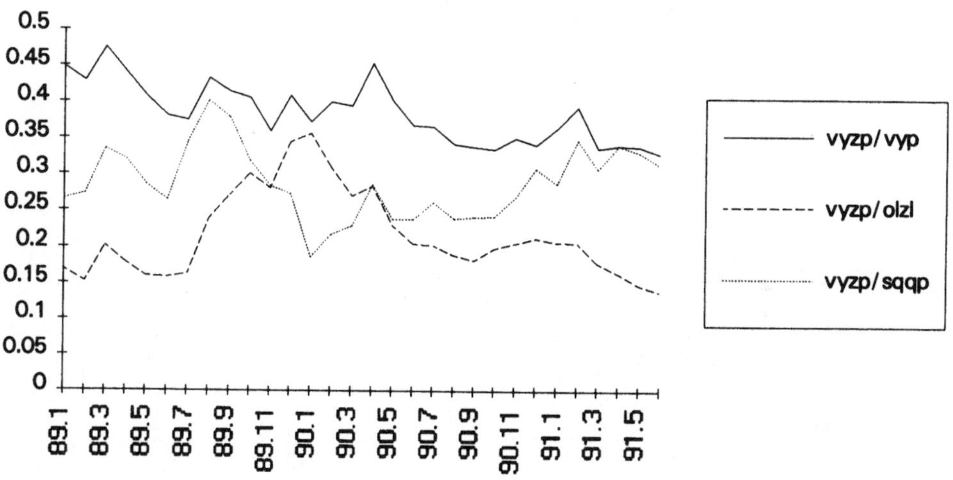

Figure 8.2:
Chosen Economic Indicators

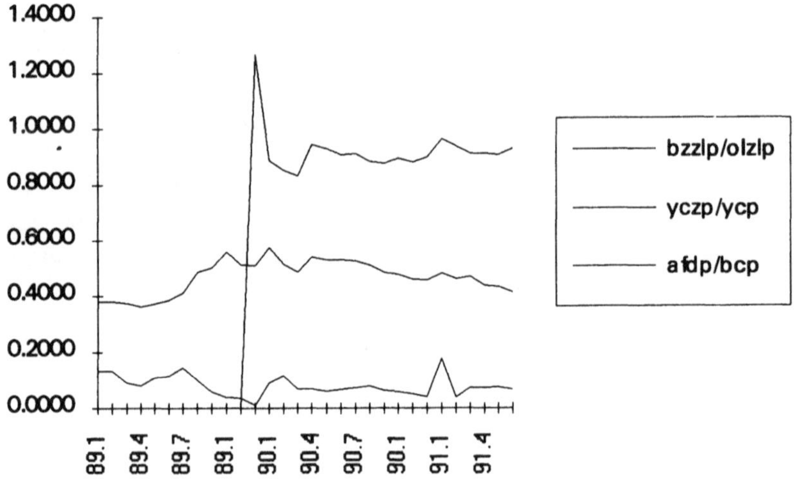

Figure 8.3:
Chosen Economic Indicators

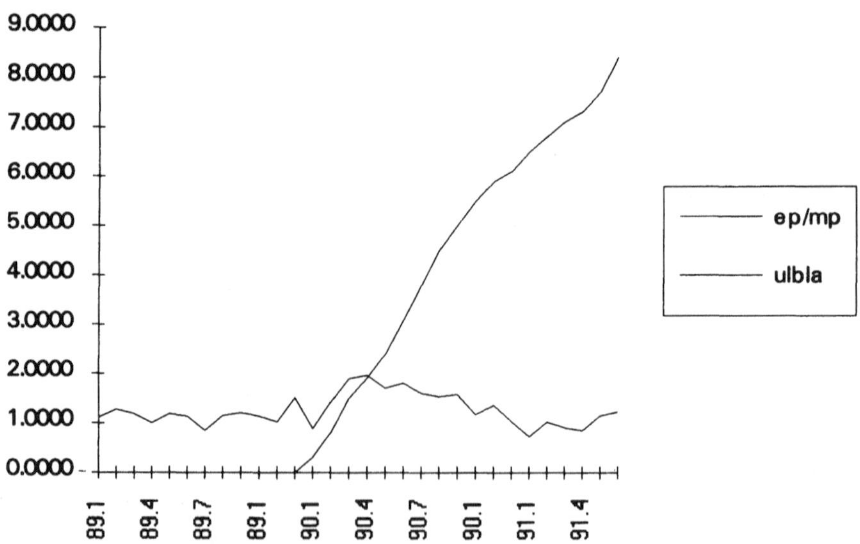

Table 8.1:
Chosen Economic Indicators for Polish Economy

Period	[1]	[2]	[3]	[4]	[5]	[6]	[7]	[8]
1989.1	0.4487	0.1692	0.2658	1.1108	0.0	0.0	0.3803	0.1340
1989.2	0.4294	0.1521	0.2726	1.2743	0.0	0.0	0.3798	0.1324
1989.3	0.4757	0.2022	0.3449	1.1735	0.0	0.0	0.3730	0.0917
1989.4	0.4409	0.1786	0.3200	1.0017	0.0	0.0	0.3615	0.0800
1989.5	0.4077	0.1600	0.2865	1.1742	0.0	0.0	0.3712	0.1100
1989.6	0.3810	0.1587	0.2643	1.1310	0.0	0.0	0.3850	0.1159
1989.7	0.3743	0.1634	0.3444	0.8471	0.0	0.0	0.4120	0.1448
1989.8	0.4326	0.2389	0.4020	1.1404	0.0	0.0	0.4850	0.1014
1989.9	0.4141	0.2712	0.3784	1.2002	0.0	0.0	0.5014	0.0581
1989.10	0.4059	0.3007	0.3181	1.1307	0.0	0.0	0.5565	0.0389
1989.11	0.3590	0.2809	0.2836	1.0202	0.0	0.0	0.5116	0.0370
1989.12	0.4087	0.3446	0.2739	1.5034	0.0	1.2651	0.5098	0.0105
1990.1	0.3725	0.3563	0.1867	0.8862	0.0079	0.8868	0.5736	0.0914
1990.2	0.4002	0.3092	0.2177	1.4120	0.0219	0.8519	0.5125	0.1156
1990.3	20.3947	0.2728	0.2294	1.8920	0.0389	0.8329	0.4846	0.0683
1990.4	0.4538	0.2829	0.2864	1.9651	0.0519	0.9427	0.5400	0.0690
1990.5	0.4028	0.2303	0.2388	1.7017	0.0668	0.9285	0.5290	0.0612
1990.6	0.3673	0.2054	0.2386	1.8067	0.0866	0.9070	0.5298	0.0677
1990.7	0.3662	0.2030	0.2621	1.5942	0.1083	0.9110	0.5252	0.0726
1990.8	0.3418	0.1983	0.2394	1.5275	0.1291	0.8825	0.5085	0.0797
1990.9	0.3383	0.1824	0.2417	1.5794	0.1478	0.8769	0.4819	0.0628
1990.10	0.3350	0.1994	0.2428	1.1672	0.1630	0.8939	0.4768	0.0580
1990.11	0.3507	0.2059	0.2692	1.3336	0.1786	0.8805	0.4584	0.0509
1990.12	0.3409	0.2129	0.3082	1.0357	0.1904	0.8986	0.4572	0.0409
1991.1	0.3647	0.2079	0.2873	0.7254	0.1875	0.9632	0.4801	0.1775
1991.2	0.3943	0.2068	0.3486	1.0238	0.1984	0.9374	0.4611	0.0384
1991.3	0.3366	0.1785	0.3079	0.9050	0.2078	0.9114	0.4690	0.0726
1991.4	0.3405	0.1647	0.3408	0.8445	0.2180	0.9117	0.4370	0.0730
1991.5	0.3388	0.1483	0.3310	1.1420	0.2317	0.9074	0.4320	0.0748
1991.6	0.3286	0.1392	0.3137	1.2229	0.2543	0.9291	0.4130	0.0679

Figure 8.2 presents indicators apparently independent from each other. The central line presents the part of expenditures on food in total expenditures of households. The increase of food prices associated with the decline of subsidies from the state budget in the second half of 1989 resulted in an increase of expenditures on food purchase to more than 50 % of total expenditures. Prices stabilization in 1990 allowed to decrease value of this indicator about 10 percent points. The bottom line shows the ratio of subsidies from state budget to total expenditures of the state budget. Its declining tendency was just indicated. The last indicator presented in this figure shows the relation between total indebtedness level of the population and the level of bank savings deposits of households. Statistical data concerning this variable are not obtained until December of 1989. The general tendency of this indicator is rather growing, however, in January and February of 1990 a deep decrease occurred. During these two months the National Bank of Poland decided, due to high inflation rates, to regulate the debt level of population by increasing interest rates on credits (40 % in January, and a bit less in February). People reacted by paying off money and used for this purpose bank savings. While observing separately two time series it is obvious that during the first months of 1990 the debt of population decreased by 40 % and starting with March it has been systematically increasing. Savings during these months globally not decreased but their growth rate was very low in comparison with other periods. First of all, deposits in hard currency have been withdrawn.

Figure 8.3 contains two indicators essential for our economy: the relation between total exports and total imports and the unemployment rate. Due to high indebtedness of our economy, exports should exceed imports. Such a favorable situation took place throughout 1990 (but at the beginning of 1991 this relation was changed) due to a decrease in exports and a simultaneous increase in imports (mainly of consumer goods).

Unemployment is a totally new phenomenon in our economy and for the first time appeared in December 1989. The unemployment rate, which increased very quickly within several months is very alarming because several new problems must be solved. Forecasting the discussed indicators for only a couple of periods is very risky. One can just hope that at least the disadvantageous tendencies will not deepen.

The next stage of the analysis comprised an investigation of the particular series by means of Brown exponential smoothing. If there were signs of seasonality, also Winters decomposition was used. Generally, those variables which represent the financial sector of our economy in the recent years were presented. For each variable there was chosen only one (the best) variant of exponential smoothing. Forecasts were calculated for 12 months. When choosing the "optimal" variant, the ex post errors of forecast were not the only guide lines. Sometimes knowledge based on the government decisions was used.

Table 8.2:
Summary Statistics and Forecasts for Variable AFDP

Type of function	M.E.	M.S.E.	M.A.E.	M.A.P.E.	M.P.E.
-61.8192+54.4251*T	0.00000	90859.5	213.687	52.5954	-26.2459
-198.558+80.0637*T-0.827052*T^2	0.00000	87798.5	214.666	58.5745	-19.0340
EXP(4.61671+0.105236*T)	22.1445	197171	299.363	39.9798	-12.0062
EXP(6.68802-3.30523/T)	204.287	246801	411.893	96.3303	-44.0631
Simple: 0.9	45.3668	160831	216.663	32.8290	-3.6900
Linear: 0.2	28.7753	129399	210.377	28.3155	-1.97273
Linear: 0.5	-2.59356	186737	245.694	37.4275	-10.0851
Linear: 0.9	0.87690	152044	384.212	56.3895	-12.7362
Quadratic: 0.2	-15.3755	151945	250.993	34.4115	-8.90562
Quadratic: 0.5	-4.10107	310759	325.424	50.0127	-13.6306
Quadratic: 0.9	8.69595	1.19942E6	637.478	89.1115	-13.9506

Results of these forecasts for a dozen of categories are presented graphically in the appendix to this paper (figures 8.A1–8.A20). Many of the selected series are characterized by strong irregularities, hence, the selection of the presented forecasting methods seems right. As an example for one the variables (AFDPM the subsidies from the state budget) a comparison of this forecasting technique with various trend forms is presented in tables 8.2 and 8.3 and figures 8.A21–8.A31. For selecting the best variant of the forecasting formula one cannot simply consider errors regarding the statistical sample. Using the formula and analyzing MPE errors one needs to choose the quadratic formula 0.9, which, however, gives unrealistic high values of the forecasts. It is impossible for the subsidies to grow at a rate of 50 % month after month. The Polish government tends to gradually eliminate price subsidizing, but, certainly, this cannot be achieved immediately. Hence, most realistically, values of the variable are extrapolated by version "simple 0.9" although the ex-post error of forecasts is not the lowest.

For the comparison's sake one would also need forecasts determined according to the causal relations. Such models are much more stable as regards their parameters. This stage of analysis can be done a little later, when equations of the model, describing relations between the elements of the national economy have been specified.

Table 8.3:
Forecasts of Variable AFDPM for the Period 7.91 – 6.92

Type	7.91	8.91	9.91	10.91	11.91	12.91
1	1679.7	1734.2	1788.6	1843.0	1897.5	1951.9
2	1516.5	1542.8	1567.5	1590.5	1611.8	1631.5
3	2934.4	3260.0	3621.8	4023.8	4470.3	4966.4
4	723.9	726.2	728.3	730.3	732.3	734.1
5	1303.1	1303.1	1303.1	1303.1	1303.1	1303.1
6	1473.1	1505.4	1537.6	1569.8	1602.0	1634.3
7	1265.3	1247.1	1229.0	1210.8	1192.7	1174.5
8	1345.0	1364.9	1384.8	1404.7	1424.6	1444.5
9	1356.9	1352.6	1345.0	1334.0	1319.8	1302.2
10	1154.6	1067.2	966.0	850.9	722.1	579.3
11	1936.7	2508.1	3250.7	4164.4	5249.3	6505.4

Type	1.92	2.92	3.92	4.92	5.92	6.92
1	2006.3	2060.7	2115.1	2169.6	2224.0	2278.4
2	1649.6	1665.9	1680.7	1693.7	1705.2	1714.9
3	5517.5	6129.8	6810.1	7565.8	8405.4	9338.2
4	735.8	737.5	739.0	740.5	741.9	743.3
5	1303.1	1303.1	1303.1	1303.1	1303.1	1303.1
6	1666.5	1698.7	1731.	1763.2	1795.4	1827.6
7	1156.4	1138.2	1120.0	1101.9	1083.7	1065.6
8	1464.4	1484.3	1504.1	1524.0	1543.9	1563.8
9	1281.3	1257.0	1229.5	1198.6	1164.4	1126.9
10	422.8	252.4	68.1	-129.9	-341.8	-567.6
11	7932.6	9531.0	11300.6	13241.3	15353	17636

APPENDIX

Figure 8.A1:

Original Series and Forecasts for Z6PM

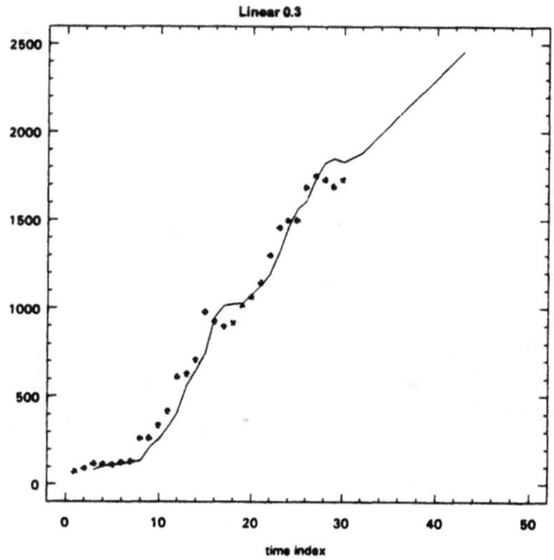

Figure 8.A2:

Original Series with Forecasts for BZUSPM

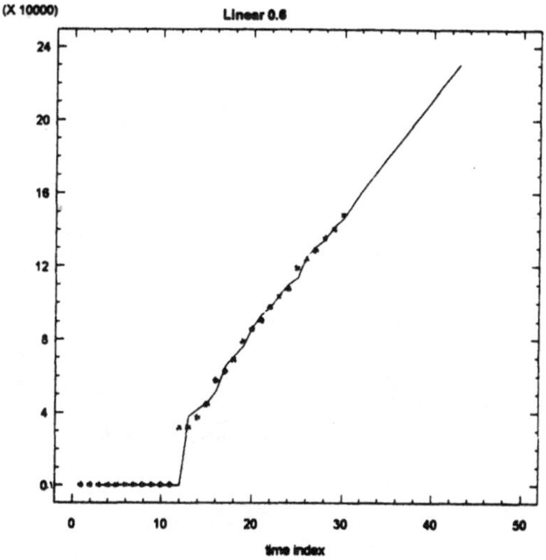

Figure 8.A3:

Original Series and Forecasts for OLPM

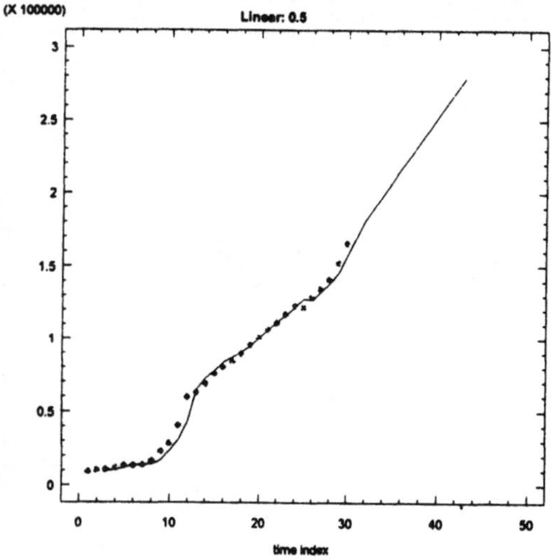

Figure 8.A4:

Original Series with Forecasts for BZLZLPM

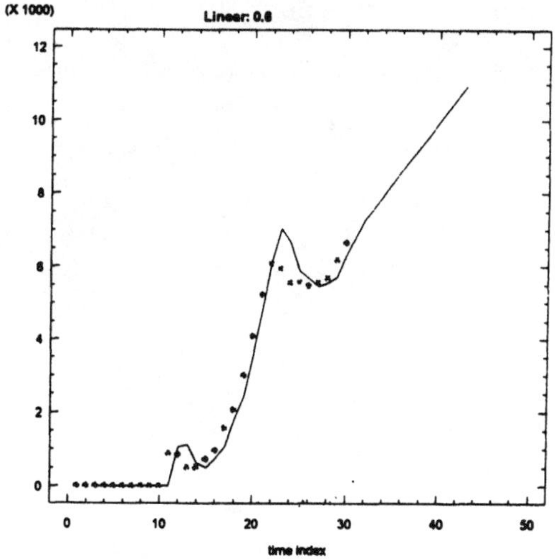

Figure 8.A5:

Original Series with Forecasts for AFDPM

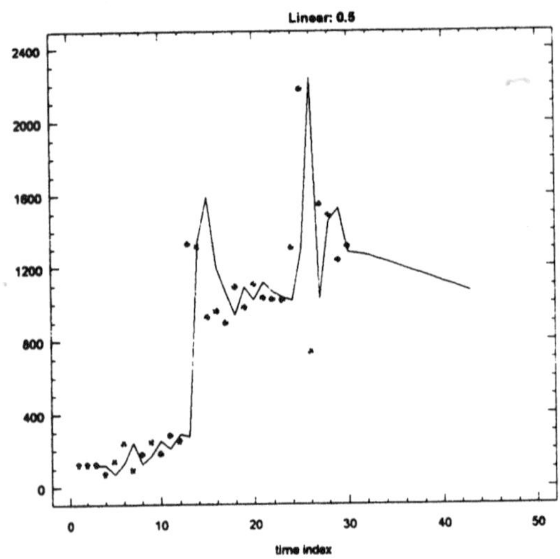

Figure 8.A6:

Original Series with Forecasts for SQQPM

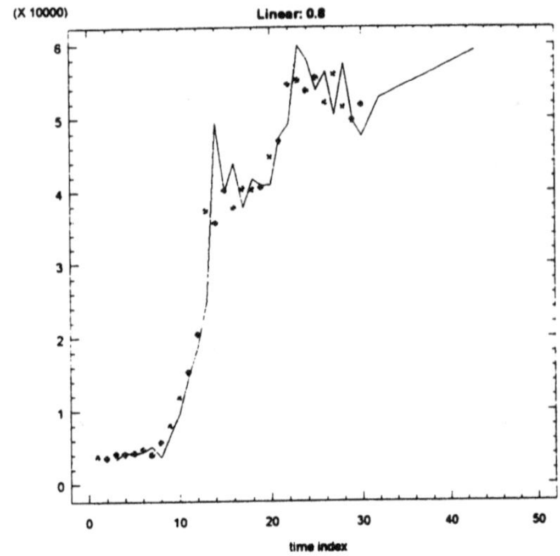

Figure 8.A7:

Original Series with Forecasts for OZLPM

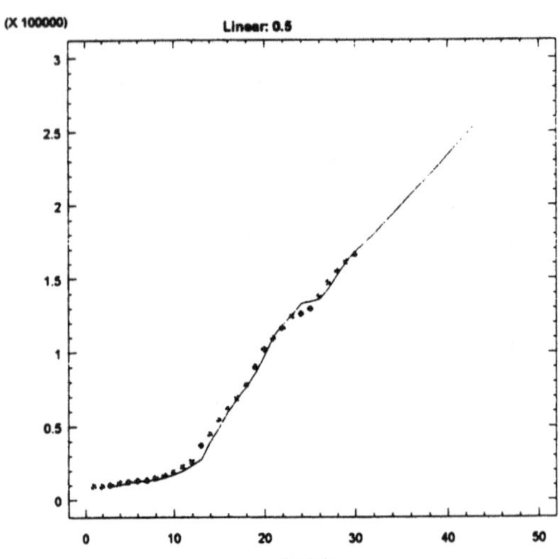

Figure 8.A8:

Original Series with Forecasts for LZBM

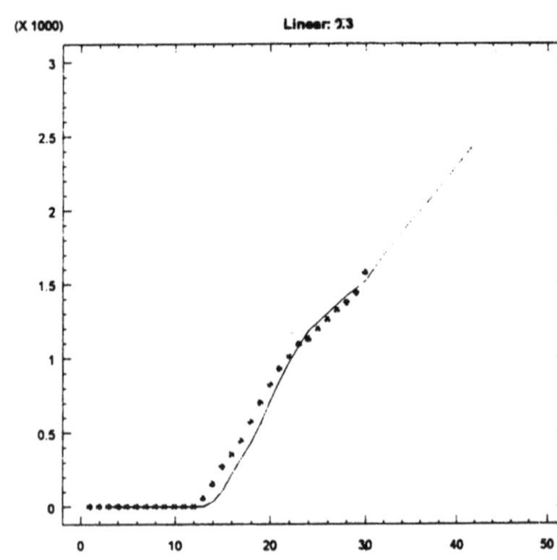

Figure 8.A9:

Original Series with Forecasts
for BYPM

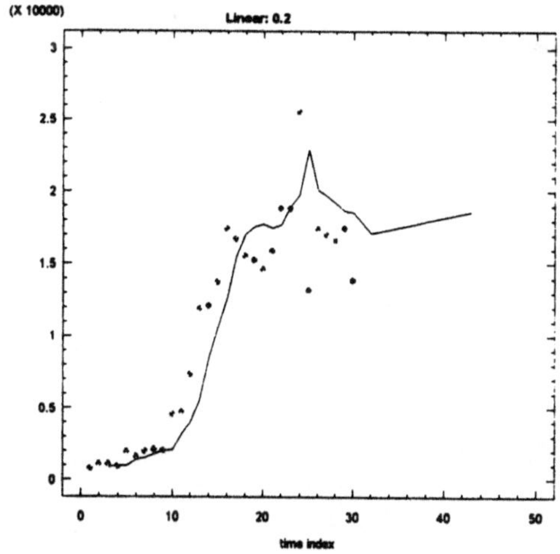

Figure 8.A10:

Original Series with Forecasts
for AFPDPM

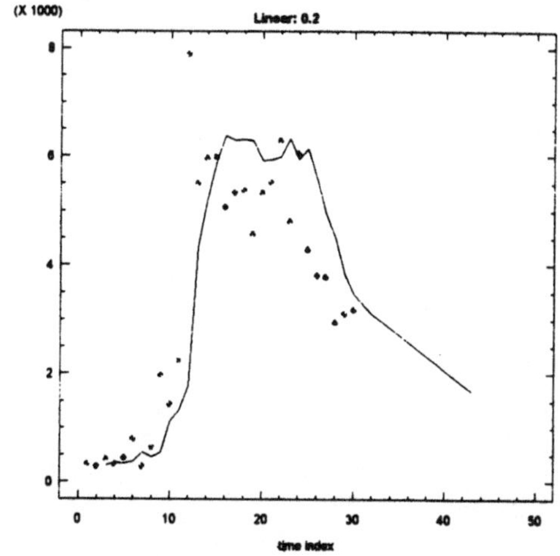

Figure 8.A11:

Original Series with Forecasts
for BCPM

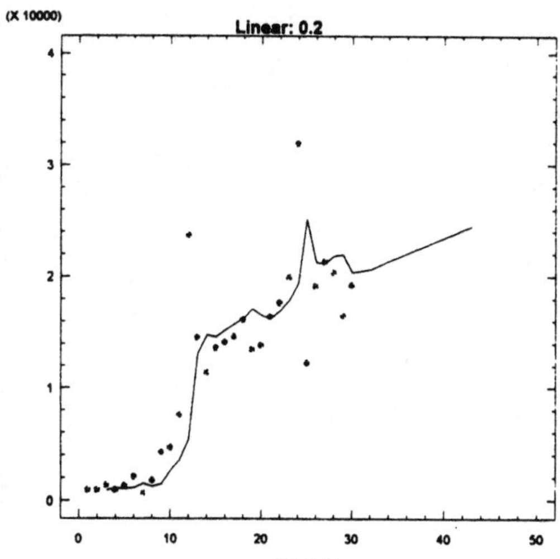

Figure 8.A12:

Original Series with Forecasts
for AFPOPM

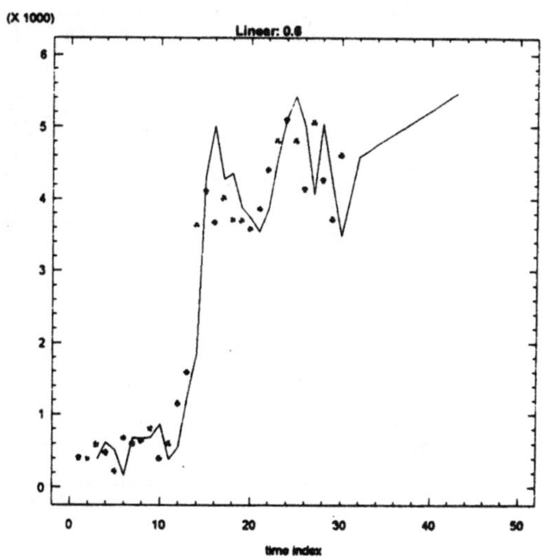

Figure 8.A13:

Original Series with Forecasts for EPM

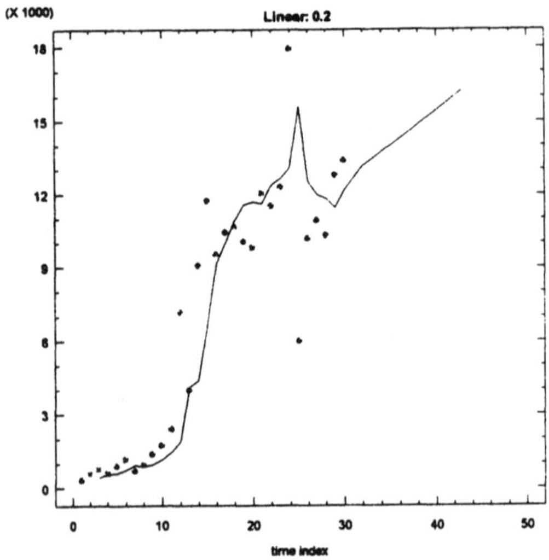

Figure 8.A14:

Original Series with Forecasts for EDPM

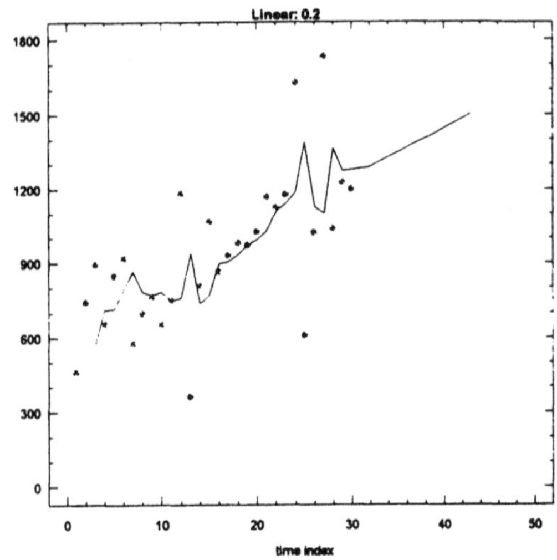

Figure 8.A15:

Original Series with Forecasts for MPM

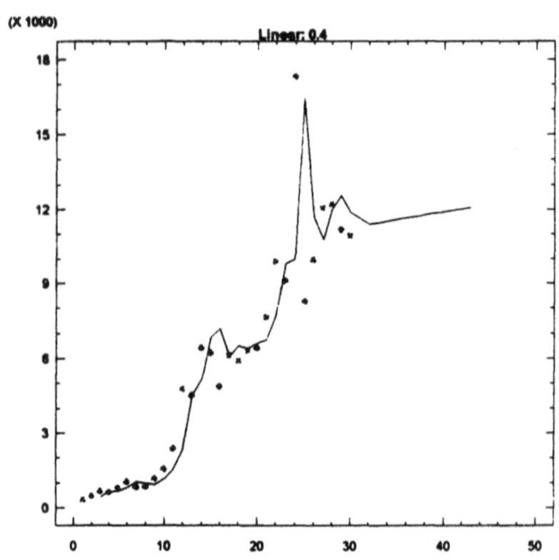

Figure 8.A16:

Original Series with Forecasts for MDPM

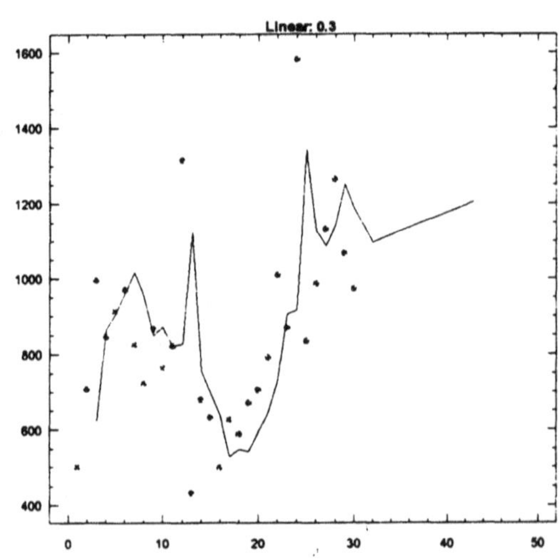

Figure 8.A17:

Original Series with Forecasts for VYPM

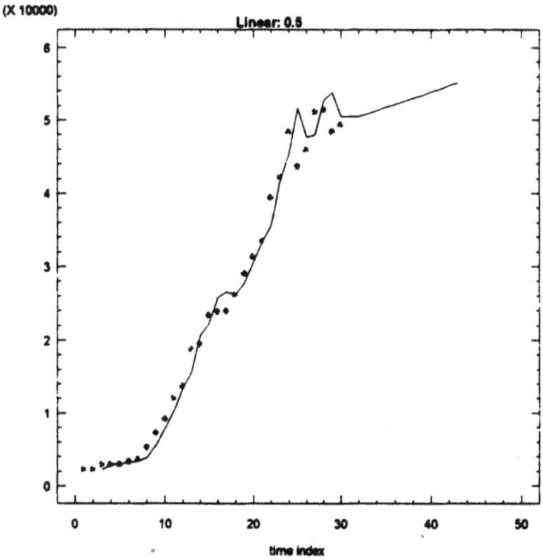

Figure 8.A18:

Original Series with Forecasts for VYZPM

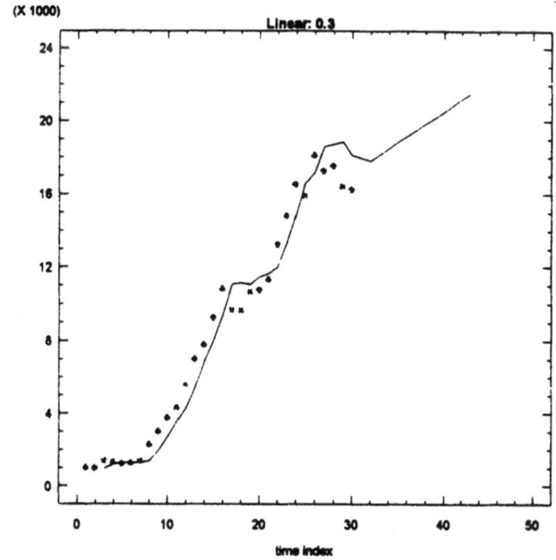

Figure 8.A19:

Original Series with Forecasts for VCPM

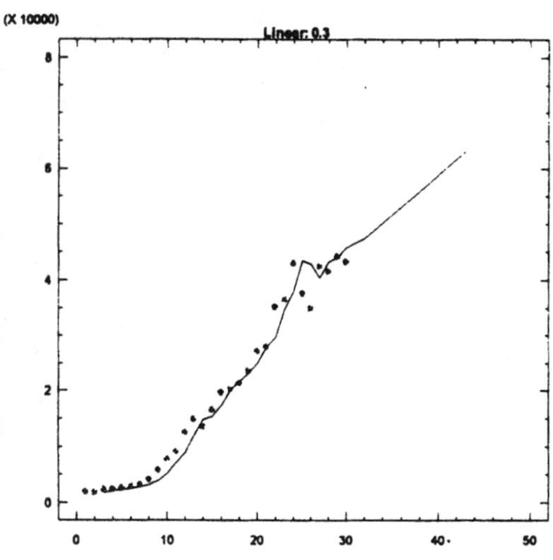

Figure 8.A20:

Original Series with Forecasts for VCHPM

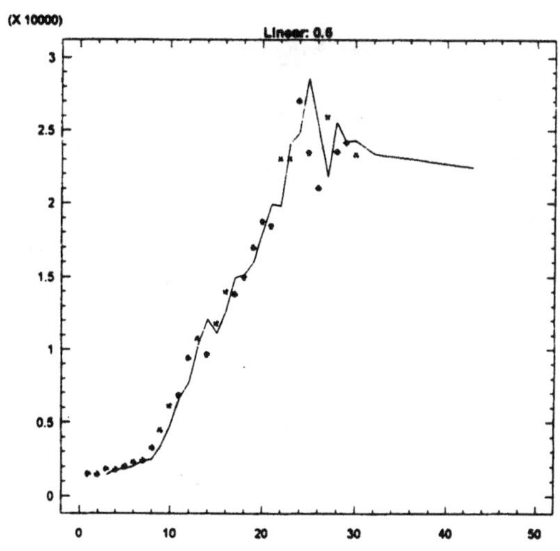

Forcasts for variable AFDP – subsidies from the state budget

Figure 8.A21:

Original Series with Forecasts
-61.8192+54.4251*T

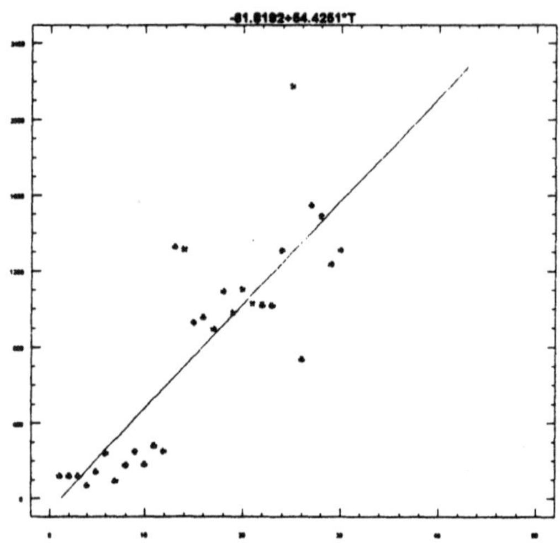

Figure 8.A22:

Original Series with Forecasts
-198.558+80.0637*T-0.827052*T^2

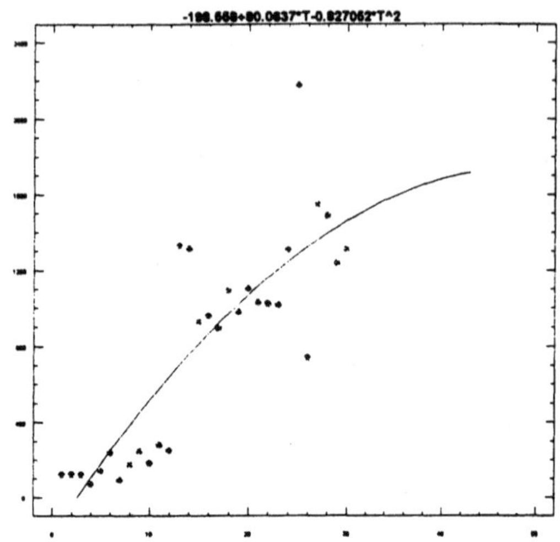

Figure 8.A23:

Original Series with Forecasts
EXP(4.61671+0.105236*T)

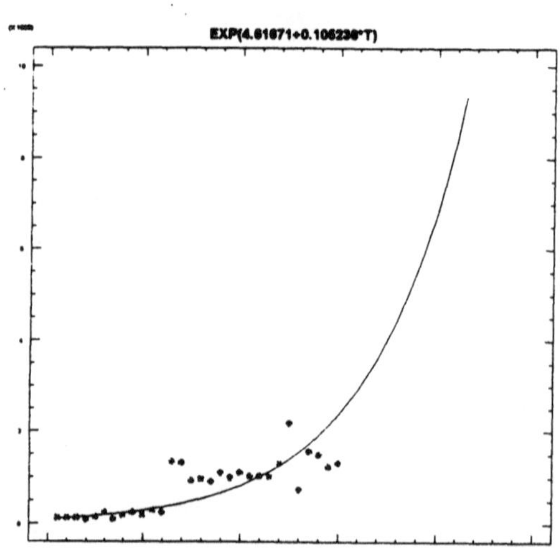

Figure 8.A24:

Original Series with Forecasts
EXP(6.68802-3.30523/T)

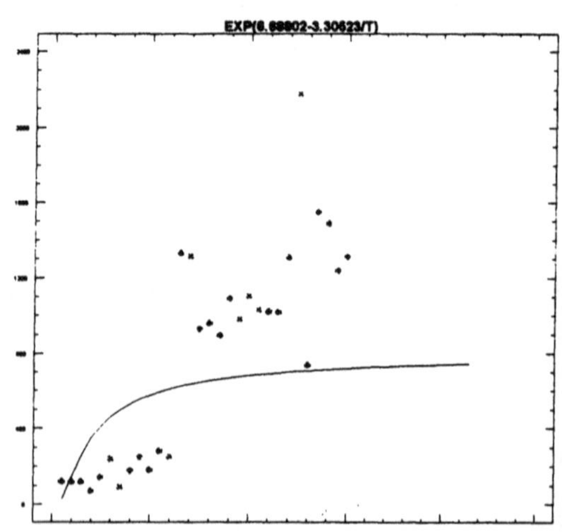

Figure 8.A25:

Original Series with Forecasts
Simple: 0.9

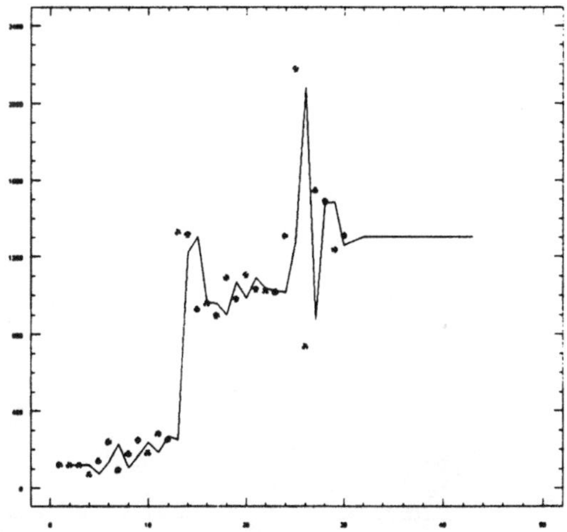

Figure 8.A26:

Original Series with Forecasts
Linear: 0.2

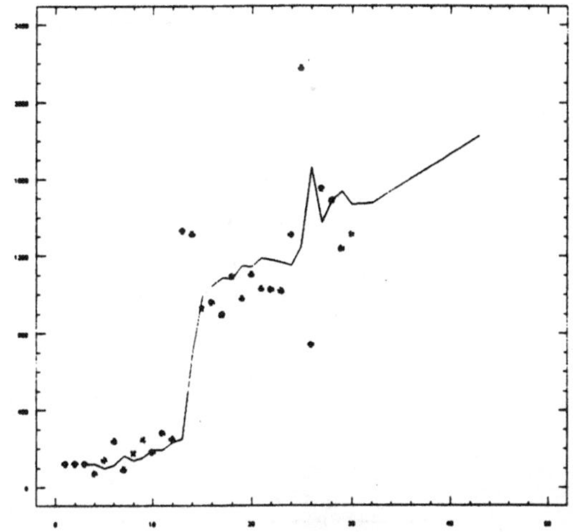

Figure 8.A27:

Original Series with Forecasts
Linear: 0.5

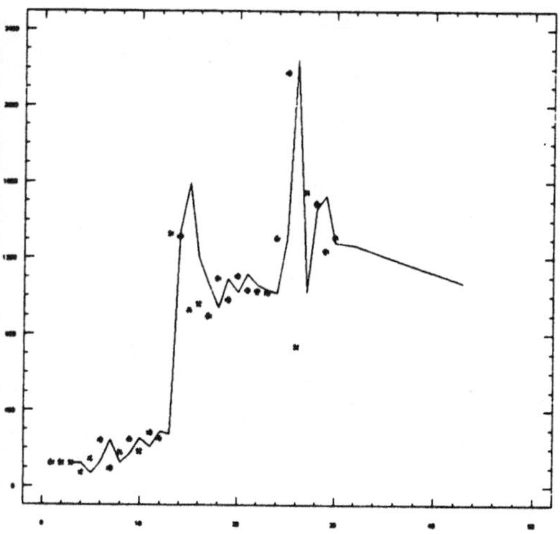

Figure 8.A28:

Original Series with Forecasts
Linear: 0.9

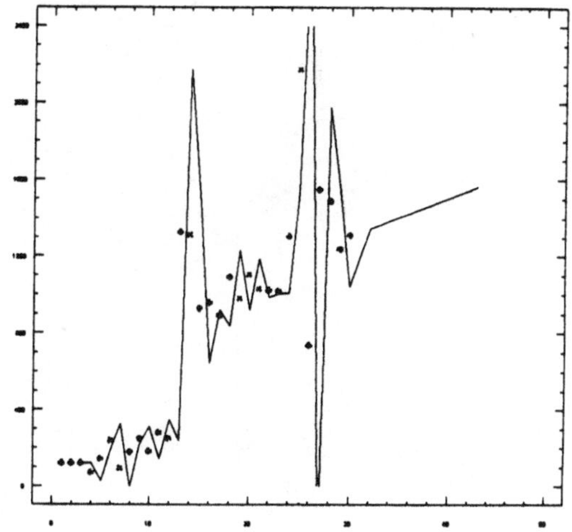

Figure 8.A29:
Original Series with Forecasts
Quadratic: 0.2

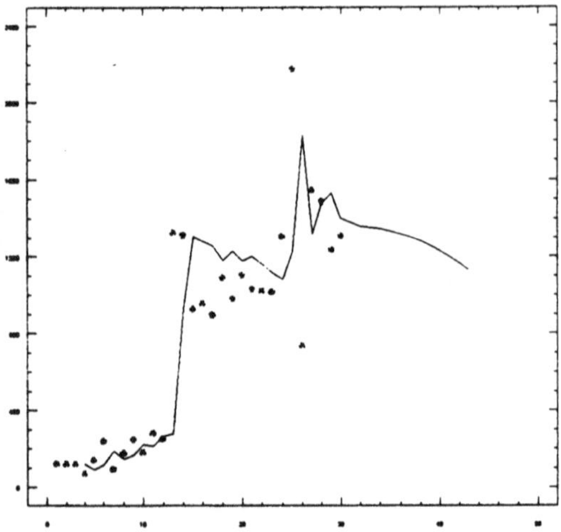

Figure 8.A30:
Original Series with Forecasts
Quadratic: 0.5

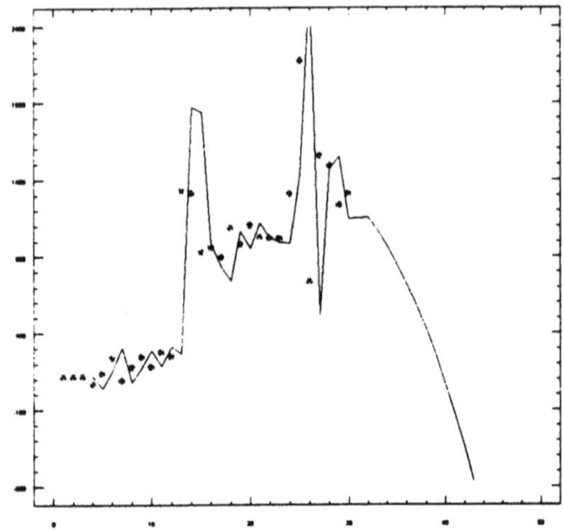

Figure 8.A31:
Original Series with Forecasts
Quadratic: 0.9

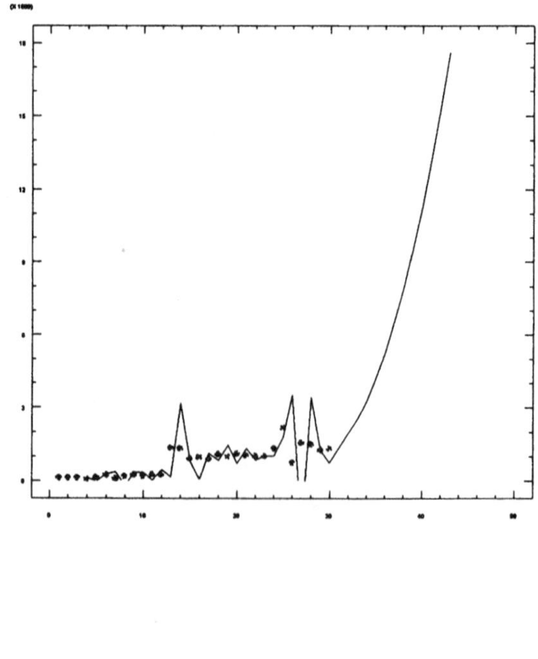

References

Abraham, B. and J. Ledolter (1983): "Statistical Methods for Forecasting." John Willey and Sons.

Box, G.E.P. and G.M. Jenkins (1976): "Time Series Analysis: Forecasting and Control." 2nd edition, Holden Day, San Francisco.

Brown R.G. (1962): "Smoothing, Forecasting and Prediction of Discrete Time Series." Practice Hall, Englewood Cliffs, NJ.

Brown, R.G. and R.F. Meyer (1961): "The Fundamental Theorem of Exponential Smoothing." *Operation Research*, **9**: 673-685.

Engle, R.F., C.W.J. Granger, S. Hylleberg, and H.S. Lee (1990): "Seasonal cointegration: The Japanese Consumption function 1961.1–1987.4." Paper presented at the Econometric Society World Congress, Barcelona, Spain.

Hellwig, Z. (1967): "Schemat budowy prognozy statystycznej metoda wag harmonicznych." Przeglad Statystyczny, nr 2.

Hylleberg, S., R.F. Engle, C.W.J. Granger, and B.S. Yoo (1990-HEGY): "Seasonal Integration and Cointegration." *Journal of Econometrics*, **44**: 215-238.

Ledolter, J. and G.E.P. Box (1978): "Conditions for the Optimality of Exponential Smoothing Forecast Procedures." *Metrica*, **25**: 77–93.

Lee, H.S. (1989): "Maximum Likelihood Inferences on Cointegration and Seasonal Cointegration." Discussion paper, University of California, San Diego.

Kolupa, M. (1964): "Prognozy popytu a autokorelacja skladnika losowego." Przeglad Statystyczny, nr 3.

Kulawik-Hlousek, K. (1972): "Dokladnśc prognoz a odleglość okresu prognozowanego od okresu bieàcego, Wiadomoći Statystyczne, nr 9.

Kunst, R.M. (1990): "Seasonal Cointegration in Macroeconomic Systems: Case Studies for Small and Large European Countries." *Research Memorandum*, No. 271, Institute for Advanced Studies, Vienna.

Marciszewska, S. (1973): "Zmodyfikowana metoda wykładniczego wyrównywania danych do budowy prognoz w warunkach nieliniowych zmian funkcji trendu, Wiadomości Statystyczne.

Pawłowski, Z. (1970): "Wyrównywanie wykładnicze w warunkach wahań periodycznych zmiennej prognozowanej." *Przeglad Statystyczny*, nr 2.

Pawłowski, Z. (1973): "Prognozy ekonometryczne." PWN, pp. 286.

Pawłowski, Z. (1974): "Teoria prognozy ekonometrycznej w gospodarce socjalistycznej." PWN, p. 242.

Podgórska, B. (1972): "Problemy wyboru jednej z wielu prognoz zbudowanych na ten sam okres zmodyfikowana metoda wyrównywania wykładniczego." *Wiadomości Statystyczne*, nr 10.

Zadora, K. (1969): "Predykcja szeregów czasowych metoda wyrównywania wykładniczego, *Przegläd Statystyczny*, nr 2.

Chapter 9

Forecasting with Short and Seasonally Unadjusted Data: The Structural Modeling Approach*

BALDEV RAJ AND THOMAS URL

1 Introduction

A researcher undertaking forecasting or prediction using either econometric, time series or exponential smoothing methods encounters a number of problems, pitfalls, and challenges, if the available data are short and/or contain irregularities such as missing observations, data revisions or temporal aggregation. Data problems of this kind are less likely to be a major obstacle for macroeconomic forecasting in most advanced economies, since sufficient data are readily available from statistical agencies of the government, and most problems outlined earlier have generally been resolved beforehand. Moreover, most forecasting methods can handle both detection and modeling of structural breaks in the data. In addition, the dating of a structural break is aided by the availability of detailed studies in economic history. The situation may not be so fortunate if forecasting has to be performed in an uncertain and frequently changing environment such as that currently experienced by several Eastern European countries. Countries in this area are attempting to make

*We wish to thank A. Jaeger, R. Kunst, A. Wörgötter, and conference participants for helpful comments.

a transition from a command economy to a market economy as rapidly as possible, taking into account both political and economic constraints. An important question faced by forecasters in government offices and firms is, what method of forecasting should be used to predict a set of relevant economic variables by using an appropriate statistically based model. In addition the method selected should provide an accurate description of key features in the variables of interest, such as trend and seasonal. It is known that the choice of a method depends on a number of factors such as the personal preferences of the forecaster, the choice of the criterion deemed appropriate to measure forecast accuracy, and the costs of the method. Another important consideration may involve taking into account the instability of relations among economic variables resulting from the transition process. These arguments suggest that the use of univariate forecasting methods as opposed to either multivariate econometric or multivariate time series methods may be more appropriate in support of a decision making process during the early stages of transition. In view of alternative theoretical arguments to be provided later on in the paper our preference is for using the Structural Time Series (STS) model (cf. Harvey [1989]) as opposed to the Box-Jenkins ARIMA model. The former class of models combines the classical decomposition of a time series into trend, seasonal, and irregular components with a parsimonious characterization of a slowly changing trend component. It introduces flexibility by allowing coefficients to change over time. In the earlier econometrics literature this type of decomposition was well known as the unobserved-component (UC) model (cf. Nerlove [1972], Engle [1978], and Nerlove et al. [1979]). The UC model has also been used by others such as Nelson and Plosser [1982] and Watson [1986] in a different but somewhat related context.

The STS class of models is closely related to Box-Jenkins ARIMA models, in the sense that ARIMA models can be interpreted as an unrestricted reduced form of the STS models. Moreover, as has been shown by Nelson [1988], the forecast functions of a particular member, called the Basic Structural Model (BSM) which we use in this paper, yields an almost similar forecasting function as the multiplicative seasonal ARIMA model. In the ensuing discussion, the linkage between these two classes of models will prove useful in interpreting empirical results of the forecasting performance. In support of our choice of the BSM vis-à-vis the multiplicative seasonal ARIMA method of forecasting, we also present a number of theoretical arguments drawn mostly from the existing literature. This account of theoretical support for the choice serves to motivate the empirical analysis to be presented. Moreover, we provide intuitive arguments instead of formal proofs since the latter type of material is already available elsewhere, such as in Harvey [1989]. A recent survey of this type of modeling is provided by Harvey and Shepard [1993].

The empirical evidence is based on 40 quarterly and 120 monthly observations on three variables each for Austria and Canada. In order to provide evidence that mimics as closely as possible the scenario faced by East European economies, we use seasonally unadjusted data without correcting for outliers. Additionally, the time period evaluated contains some turbulences. For instance some of the variables experienced a possible change in either level or slope or a combination of both. Furthermore, four of the series are aggregated by picking out end of quarter values from a monthly series. This results in a loss of information due to larger measurement intervals compared to decision making intervalls of economic aspects and can be considered normal in the context of a changing economic environment. Although the data come from two advanced economies, these features should provide a flavor of the relative performance of the two competing methods of forecasting in an environment similar to that likely to occur in an East European country's situation.

The motivation for providing some first-hand experience in situations as described above comes from a different source. The earlier empirical evidence in support of the view that the STS model has a better forecasting performance relative to unrestricted time series ARIMA methods as presented in Harvey and Todd [1983] has recently been the subject of some dispute. For instance, it has been argued by Garcia-Ferrer and Del Hoyo [1992] that the forecasting performance of ARIMA models is superior to that from STS models, including the BSM considered in this paper. Their evaluation, however, is based on a longer time horizon of 35 years as opposed to the short time horizon of 10 years used in this paper. In summary the evidence in the literature can be seen as less than firmly in support of the STS model relative to the ARIMA model. It needs to be stressed that the relative performance of these two methods to some extent depends on the choice of the criterion used to evaluate the accuracy of the forecast as observed by Nerlove et al. [1979] among others. A novelty of the evidence to be presented by us is that besides using standard criteria such as the mean squared errors of forecasts we provide evidence of the relative forecast efficiency based on a test of forecast encompassing. This test gives useful information about the question of whether the forecast from one method can be improved by incorporating information from the second method. While the use of forecasting encompassing tests is not new in comparing forecasts of competing methods, we believe its use in this particular context is new. To anticipate the conclusions of our empirical investigations, we find that forecasts from neither method encompass the other at the 5 % significance level. In other words the BSM produces forecasts that are at least as good as those of the ARIMA model.

The rest of the paper is organized as follows: In section 2, the specification of the Basic Structural Model is presented. This section also contains the reformulation of the model in the State Space form, estimation of the unknown parameters of the model, and predictions of future observations together with their conditional mean square errors based on a Maximum Likelihood estimation method. Section 3 gives a brief discussion of the model evaluation criteria and the results of modeling a small selection of macroeconomic time series from Austria and Canada. The data used are complex enough that a decade of quarterly observations on variables would be considered short by most definitions. Section 4 concludes the paper.

2 The Basic Structural Model

The specification of the Basic Structural Model (BSM) utilizes the intuitive idea that most macroeconomic time series could be viewed as being comprised of three components: trend, seasonal and irregular. This idea is then modified to accommodate an other intuitively appealing idea that a sensible method of forecasting should place more weight on recent observations compared to distant past observations. More formally, the structural modeling approach is based on the postulate that the trend in the time series of interest is stochastic, such that it evolves slowly over time according to a random walk process. In the same vein it is postulated that the seasonal component in the series is stochastic, such that it also evolves relatively smoothly over time around an average zero level with constant variance. Finally, the irregular component is assumed to be a stochastic white noise process.

The discussion above may have made clear that the primary motivation of the BSM is to keep specifications of various components in the series simple and parsimonious so as to mimic the main features of most seasonal economic time series. The BSM, which is a particular member of the class of structural models, can be generalized to allow for a more complex stochastic approximation, such as a polynomial trend, a more elaborate stochastic seasonal component, such as a set of trigonometric functions at seasonal frequencies instead of the dummy variable method, and a more elaborate irregular component which could be modeled as a stationary ARMA process. Finally, a cycle component can also be brought into the model for fairly long time series. However, for short time series these generalizations may unnecessarily complicate the structure, making the model less parsimonious.

All structural time series models including the BSM focus on a simple but sensible specification of various components in the time series which have direct interpretations; as argued previously this feature can be appealing for the specification of the model at initial stages. It also curtails ambiguity in the interpretation of the reduced form by imposing a parametric structure on the series. The BSM can also be viewed as a sum of three Box-Jenkins ARIMA models that are not directly observed. In this sense the BSM belongs to a class of unobserved components ARIMA (UCARIMA) models (e.g. Engle [1978], Nerlove et al. [1979]). The BSM is similar to the Holt-Winters additive exponential smoothing forecasting method, where smoothing parameters perform a role similar to variances of the disturbance terms in the BSM relative to the variance of the irregular component. Moreover, the BSM can be viewed as a close relative of Bayesian forecasting models introduced by Harrison and Stevens [1971, 1976], although the BSM is grounded in the strictly classical statistics traditions. The important distinction between the BSM and Bayesian methodologies is that in the former models the variances of the disturbance terms are regarded as fixed parameters to be estimated, while in the latter models these parameters are allowed to shift in response to a structural change in the process generating the data at unknown points. This linkage is important since Bayesian forecasting models provide another class of successful methods for forecasting short time series, especially those involving a structural change in the data generating process (e.g. Fildes [1983, 1984]).

2.1 Model Specification: The Basic Structural Model

Let y_t represent the tth observation on the variable to be forecasted. Then the BSM is comprised of trend, seasonal and residual components of the form:

$$y_t = p_t + s_t + u_t, \qquad t = 1, 2, ..., T, \tag{9.1}$$

where p_t, s_t and u_t are respectively the stochastic trend, seasonal dummy variable, and irregular components in the series. Although various components enter the model in an additive fashion, a multiplicative formulation can be obtained if the variable is in logarithms. It is postulated that the stochastic process generating the trend is of the form

$$\Delta p_t = \beta_{t-1} + a_{1t}, \qquad t = 1, 2, ..., T, \tag{9.2a}$$

$$\Delta \beta_t = a_{2t}, \tag{9.2b}$$

where β_t is the slope of the local (or stochastic) trend, $\Delta = 1 - L$ is the first-difference operator with L as the lag operator $Lp_t = p_{t-1}$, and the random variables a_{1t} and a_{2t} are assumed to be independently, normally distributed disturbance terms with zero mean and variances σ_1^2 and σ_2^2, respectively.

The seasonal component is assumed to be a stochastic formulation derived from the standard dummy variable method. Specifically, it has the form:

$$S(L)s_t = a_{3t}, \qquad t = 1, 2, \ldots, T \tag{9.3}$$

where $S(L)$ is the seasonal operator, $S(L) = \sum_{j=0}^{s-1} L^j$, s is the number of seasons in the year and the disturbance term a_{3t} is also independently normally distributed with zero mean and constant variance σ_3^2. In other words, it is postulated that the seasonal pattern is slowly changing – an assumption that may be adequate in small sample situations. Finally, the irregular component is assumed to be a white noise normal process such that $u_t \sim NID(0, \sigma^2)$.

As was pointed out earlier the BSM belongs to the class of UCARIMA models studied in the literature by Engle [1978], and Nerlove et al. [1979]. The link between the BSM and UCARIMA can be made more explicit by rewriting the BSM as:

$$y_t = \frac{a_{4t}}{\Delta^2} + \frac{a_{3t}}{S(L)} + \epsilon_t, \qquad t = 1, 2, \ldots, T, \tag{9.4}$$

where $\Delta^2 = (1 - L)^2$ and $a_{4t} = a_{1t} - a_{1,t-1} + a_{2,t-1}$. An important feature of the BSM in this form is that the trend component a_{4t}/Δ^2 and the seasonal component $a_{3t}/S(L)$ do not have a root in common; accordingly, changes in the trend pattern are not compounded with changes in the seasonal pattern. The historical background to the literature on the development of the BSM Model is conveniently provided by Harvey ([1989], pp. 22–23).

2.2 The State Space Form of the BSM

Given there are four seasons in the year (i.e. $s = 4$), the BSM for a time series y_t can be compactly written as

$$y_t = \boldsymbol{z}'\boldsymbol{\delta}_t + u_t \tag{9.5}$$

where $\boldsymbol{z} = (1, 1, 1, 0, 0)', \boldsymbol{\delta}_t = (p_t, \beta_t, s_t, s_{t-1}, s_{t-2})'$. The disturbance term u_t representing the irregular component is assumed to be normally and independently distributed with mean zero and constant variance σ^2. The elements of the state vector δ_t are not observable; however, they are assumed to be generated by the first-order Markov process:

$$\boldsymbol{\delta}_t = \boldsymbol{C}\boldsymbol{\delta}_{t-1} + \epsilon_t \tag{9.6}$$

where the matrix $\boldsymbol{C}$ is given by

$$\boldsymbol{C} = \begin{bmatrix} 1 & 1 & 0 & 0 & 0 \\ 0 & 1 & 0 & 0 & 0 \\ 0 & 0 & -1 & -1 & -1 \\ 0 & 0 & 1 & 0 & 0 \\ 0 & 0 & 0 & 1 & 0 \end{bmatrix}, \tag{9.7}$$

The disturbance vector $\boldsymbol{\epsilon}_t = (a_{1t}, a_{2t}, a_{3t}, 0, 0)'$ is a sequence of normally distributed independent random variables with mean zero. The covariance of the disturbance vector is $\sigma^2 \boldsymbol{Q}$, where $\boldsymbol{Q} = diag(\sigma_1^2/\sigma^2, \sigma_2^2/\sigma^2, \sigma_3^2/\sigma^2, 0, 0)$.

The representation of the BSM in the above compact form is referred to as the State Space Form (SSF) of the model. Moreover, equations (9.5) and (9.6) are respectively termed as the measurement and the transition equations of the SSF. The SSF of the BSM makes explicit that the state vector $\boldsymbol{\delta}_t$ is nonstationary with two real roots at ± 1 and a pair of complex roots. Therefore, starting values are required for any efficient method of estimating the state vector and the unknown variances of the disturbance terms in the model.

Two further assumptions are needed to complete the specification of the SSF of the BSM. These include: (i) the initial vector $\boldsymbol{\delta}_0$ has mean $E(\boldsymbol{\delta}_0) = \boldsymbol{d}_0$ and $Var(\boldsymbol{\delta}_0) = \boldsymbol{P}_0$ and (ii) the disturbance terms u_t in the measurement equation and $\boldsymbol{\epsilon}_t$ in the transition equation are uncorrelated over all time periods and uncorrelated with the initial value $\boldsymbol{\delta}_0$.

The SSF of the BSM is time-invariant; in other words its system matrices are independent of time. This property of the model means that the BSM is always observable; however, the BSM is not controllable and cannot be stabilized unless the variances σ_2^2 and σ_3^2 are strictly positive. These concepts are related to identification conditions for the BSM.

The reformulation of the BSM in the SSF paves the way for application of the Kalman filter algorithm, which is a recursive procedure for computing the optimal estimator of the state vector, $\boldsymbol{\delta}$ at time t, conditional on the information available on y at time $t-1$ together with initial values of $\boldsymbol{d}_0$ and $\boldsymbol{P}_0$. Specifically, the optimal estimator of the state vector $\boldsymbol{\delta}_t$, based on the information up to and including time $t-1$ and the mean square error of the optimal estimator, can be obtained from the 'prediction equation' of the Kalman filter. This optimal estimator and its mean square error can also be updated by the 'updating equations' of the Kalman filter once the value of the variable y becomes available at time t (e.g. Harvey [1984], section 1). In addition, the use of the Kalman filter opens the way to obtain Maximum Likelihood estimates of the unknown parameters of the BSM model (which are the variances of the various disturbance terms) via the prediction error decomposition (see for example Schweppe [1965]; Harvey and Phillips [1979]). This framework is quite useful to modify in a straightforward manner the model to handle different irregularities in the data. This last feature may have a particular appeal for the analysis of Eastern Europe's economies.

2.3 Estimation of the Parameters of BSM

The assumption of normally distributed disturbances in the state space model implies that the Kalman filter algorithm yields the mean $\boldsymbol{\delta}_t$ and the covariance matrix $\boldsymbol{P}_t$ of the multivariate normal distribution of the state vector conditional on information at time $t-1$. Furthermore, the conditional mean of $\boldsymbol{\delta}_t$ can be regarded as an estimator of the state vector associated with a particular realization of the variable. This estimator, which is the minimum mean square error estimator of the state vector, is known to be unconditionally unbiased, i.e. the expectation of the estimation error is zero over all variables.

Assuming the first $k = s+1$ observations can be used to construct the starting values for the Kalman filter according to the method described in Harvey ([1989], pp. 120–125)[1], the likelihood function can be formulated for $y_{k+1}, y_{k+2}, \ldots, y_T$ in terms of the prediction error decomposition; that is,

$$logL = -\frac{T-k}{2} log 2\pi - \frac{1}{2} log\sigma^2 - \frac{1}{2}\sum_{t=k+1}^{T} log f_t - \frac{1}{2\sigma^2}\sum_{t=k+1}^{T}\frac{v_t^2}{f_t}, \qquad (9.8)$$

where $v_t = y_t - \boldsymbol{z}'\boldsymbol{\delta}_{t|t-1}$ for $t = k+1, \ldots, T$ are the prediction errors and $\sigma^2 f_t$ is the variance of v_t with $f_t = \boldsymbol{z}'\boldsymbol{P}_{t-1}\boldsymbol{z} + 1$.

Numerical optimization of the likelihood function can be more conveniently carried out by working with the concentrated likelihood function, which is obtained by concentrating out σ^2 from (9.8). The main advantage of working with the concentrated likelihood function is that numerical optimization needs to be carried out with respect to one less parameter (e.g. Harvey [1989], p. 125). Then, the maximum likelihood estimator of σ^2 is given as:

$$\tilde{\sigma}^2 = \frac{1}{T-k}\sum_{t=k+1}^{T}\frac{v_t^2}{f_t} \qquad (9.9)$$

[1]Another way of computing the likelihood is to use the initial values of $\boldsymbol{\delta}_t$ and $\boldsymbol{P}_t$ for the Kalman filter such that $\boldsymbol{\delta}_0 = 0$ and $\boldsymbol{P}_0 = \kappa \boldsymbol{I}$ where κ is set to a large but finite number.

2.4 Prediction of Future Observations

The Kalman filter yields the estimate of the state vector at time T, as well as the MSE of $\boldsymbol{\delta}_T$, based on all observations. These values can be used to obtain a one-step-ahead optimal prediction of the variable

$$\tilde{y}_{T+m|T} = \boldsymbol{z}'\boldsymbol{\delta}_{T+m|T}. \tag{9.10}$$

The MSE of $\tilde{y}_{T+m|T}$ can be obtained from

$$MSE(\tilde{y}_{Tm|T}) = (\boldsymbol{z}' \cdot \boldsymbol{P}_{T+m|T} \cdot \boldsymbol{z})\sigma^2. \tag{9.11}$$

3 Evaluating Forecasting Performance

The quality of a model can be evaluated in many ways. We chose criteria based on in-sample and out-of-sample measures. The variance of the one-step-ahead prediction error provides a measure for the goodness of fit within the sample period used for the estimation of the model by a least squares procedure. For the BSM the variance of one-step-ahead prediction errors can be obtained as (e.g. Harvey [1989], p. 264)

$$\tilde{\sigma}_p^2 = \tilde{\sigma}^2 \bar{f}, \tag{9.12}$$

where $\tilde{\sigma}^2$ is the ML estimator of σ^2 and $\bar{f}$ is the steady state value of f_t, the variance of the prediction errors. The value of $\bar{f}$ can, however, be approximated by f_T, although these concepts are distinct in some respects. The prediction error variance for the BSM will be compared with an ARIMA model's prediction error variance, which can be obtained directly by the ML estimator of the variance of the disturbance term.

For our purpose the following two measures are of particular interest. An evaluation of the forecasting accuracy of rival models can be carried out by comparing the sum of squared prediction errors in the post sample period. This can be done either by using one-step-ahead predictions or multi-step-ahead forecasts of the time series under consideration. The former represent a formal statistical test of the adequacy of a model and contain all relevant information for a post sample predictive test, whereas the latter requires caution. Multi-step-ahead or unconditional forecasts are made from the same origin, thus successive forecast errors are necessarily highly correlated and give a wrong impression of the consistency of discrepancies (Box and Tiao [1976]). This can easily be seen by expressing errors from unconditional forecasts as a nonsingular transformation of one-step-ahead prediction errors. Because of this nonsingular transformation property, unconditional forecasts are autocorrelated and heteroscedastic. Any appropriate test has to correct for that fact and ends up with testing again one-step-ahead prediction errors. Nevertheless, unconditional forecasts from time series models do not suffer from a prespecified path for exogenous variables, thus they show the characteristics of the forecast function and allow for an investigation on the data admissibility of the model.

Univariate forecasts always rely on the assumption that the data generating mechanism is constant, therefore a model should be able to capture substantial deviations of future realizations from the previous path of the series. Such a deviation can be interpreted as a structural break, which can be detected by a comparison of prediction errors out-of-sample with the ones inside. A more important interpretation of such a test statistic affects the fact that excessive data mining is usually associated with the cycle of identification, estimation and diagnostic checking for ARIMA models. If a structural break can be rejected for the BSM but cannot be rejected for the respective ARIMA model, this indicates data mining and inadequacy of the chosen ARIMA model for forecasting purposes. For the BSM a test on a structural break between estimation period and a prediction horizon of length l is available by calculating the following generalized Chow test (e.g. Harvey [1989], p. 271)

$$\xi = \sum_{t=T+1}^{T+l} = \frac{v_t^2}{l\sigma_p^2} \tag{9.13}$$

where v_t^2 are squared one-step-ahead prediction errors. The sum of these errors is compared with the prediction error variance weighted by l. A similar statistic for ARIMA models can be calculated by replacing σ_p^2 with the ML estimator of the variance of the disturbances from the ARIMA model (e.g. Box and Tiao [1976]).

3.1 The Data

To illustrate the forecasting performance of the BSM we use three policy relevant, seasonally unadjusted macroeconomic series from Austria and Canada respectively. These are total employment, the consumer price index including all items and the industrial production index. A detailed description of the data sources is provided in the appendix. With the exception of Austrian employment data, which are taken from the GEN data base of the Austrian

Institute for Economic Research all series are obtained from the OECD's Main Economic Indicators. The series are available at monthly intervals, however, we aggregate both price and production indexes by selecting the end of quarter value. The intention of this aggregation is to simulate the impact of a difference between points of measurement and information already available to economic agents. Due to the structural construction of the BSM one would expect a better performance of this method concerning this type of intervention.

Time plots for logs of the variables are given in figures 9.1 to 9.6. Both employment and industrial production series show a downward movement at the beginning of the 1980's with a strong recovery after 1982. The seasonal pattern of those series is pronounced and obviously changes over time. The price indexes evolve comparatively smooth over time with a break in the slope at the beginning (Canada) and the mid 80's (Austria) respectively. Although, seasonality in the price indexes does not appear at first glance, it can be detected by analyzing the autocorrelation function of first differences. Furthermore, the indexes include all items, thus they should reflect seasonal supply and demand behavior on theoretical grounds. Both, the BSM and ARIMA models are fitted after taking logarithms of the series in order to avoid heteroscedasticity. The estimation period covers data from 1980 through 1989 leaving 40 observations for quarterly and 120 for monthly series. More data are available up to the third quarter of 1991, which are used as a yardstick to compute the out-of-sample characteristics of the estimated models.

3.2 The Results

The BSM's are estimated using the software package STAMP 3.0 with the option for exact Maximum Likelihood estimation. For all models the algorithm converges strongly. This underlines the straightforward modeling strategy implied by STS models and justifies the lack of a more sophisticated identification procedure for ARIMA models. For the estimation of ARIMA $(p, d, q) \times (P, D, Q)_s$ models we use RATS 3.11 and follow the well known three step procedure of identification, estimation and diagnostic checking. After analyzing the correlogram and partial sample autocorrelation function for levels, first differences, seasonal differences and first cum seasonal differences we decide to set d=D=1 for all series. In the next step a bias corrected form of the Akaike criterion (e.g. Brockwell and Davis [1991], p. 287) is used to determine the exact order of autoregressive and moving average polynomials. This criterion combines the goodness of fit of an estimated model with a penalty for additional parameters and enforces parsimony during the selection process.

The last step involves some diagnostic checking by Box-Ljung Q-statistics and a cumulative periodogram test recommended in Box and Jenkins [1976]. All in all, the construction of ARIMA models is associated with a bulk of investigations and personal judgements before arriving at a suitable forecasting model. This fact is also stressed in a study by Lewandowsky [1984], who classifies ARIMA models as one of the most complex univariate forecasting methods.

The estimated parameters of the BSM's and the preferred multiplicative seasonal ARIMA $(p, d, q) \times (P, D, Q)_s$ specifications for each of the six series are given in table 9.1. For the BSM of Austrian industrial production and the Canadian price index a constant level is estimated. Also the slope estimated for Austria's employment and Canada's production is constant. Despite the irregular seasonal pattern, the variance of the seasonal component σ_3^2 is estimated to be zero for half the series. It is also relatively small for the remaining ones. The variance of the irregular component is zero for both employment series, the Austrian consumer price index and Canada's industrial production. A relatively high variance of the irregular component is estimated for Austria's industrial production.

Table 9.1:
Estimated Parameters of Basic Structural and ARIMA Models for the Period 1980 to 1989

	BSM*				ARIMA model**	
series	$\sigma^2(1)$	$\sigma^2(2)$	$\sigma^2(3)$	σ^2		Q-stat***
EMPLC	.260	.001	.002	.0	$y_t = (1 - .15L + .26L^2)(1 - .82L^{12})u_t$	0.55
EMPLA	.037	.0	.004	.0	$y_t = (1 - .22L + .18L^2)(1 - .28L^{12})u_t$	0.68
CPICQ	.0	.106	.0	.007	$y_t = (1 + .38L + .52L^2)u_t$	0.56
CPIAQ	.174	.026	.008	.0	$y_t = (1 - .38^4)u_t$	0.66
IPCQ	6.000	.0	.0	.0	$y_t = (1 + .29L + .37L^2)(1 - .54L^4)u_t$	0.64
IPAQ	.0	.356	.0	3.346	$y_t = (1 - .37^4)u_t$	0.59

* all values have to be multiplied by E-04
** y_t are first cum seasonal differenced, i.e. (1-L)(1-L^s)
*** significance level of Box-Ljung Q-statistic

The estimates for multiplicative seasonal ARIMA models are also given in table 9.1. All models are first cum seasonal differenced and include only significant moving average parameters. For none of the series we choose the airline model. With the exception of Canada's consumer prices, all models need a seasonal moving average parameter to capture remaining autocorrelation after seasonal differencing. The regular moving average polynomial is either of order 2 (all series from Canada and the Austrian employment series) or 0 (Austria's price index and industrial production).

Measures of the forecasting performance are given in tables 9.2 to 9.4. Table 9.2 presents the prediction error variance for the estimation period 1980 to 1989. For both monthly collected series the measure for goodness of fit is almost the same for BSM and ARIMA models. Aggregation by dropping some information from the series results in a clear improvement of the in sample fit of the BSM. The prediction error variance of the BSM for the Canadian price series is approximately half that from the ARIMA model. The other variables are modeled by the BSM with 10 to 40 percent more accuracy. This result is remarkable since we did not undertake any efforts to improve the fit of Basic Structural Models.

Table 9.2:
Prediction Error Variances for Estimation Period 1980 to 1989*

series	ARIMA	BSM
EMPLC	0.0316	0.0312
EMPLA	0.0060	0.0072
CPICQ	0.0214	0.0141
CPIAQ	0.0424	0.0390
IPCQ	0.7141	0.6015
IPAQ	1.0640	0.7585

* all values have to be multiplied by E-03

The next step is to calculate short term predictions for all series over different horizons. Predictions of quarterly series were carried out from the first quarter of 1990 (90Q1) to the third quarter of 1991. The monthly collected series are compared by forecasts of the last month of a quarter starting with (90M3), i.e. 3,6,9 and 12th month respectively. The sum of squared errors from one-step-ahead predictions are given in tables 9.3a and 9.3b. Over this interval the results seem to depend on the type of the series. Smoothly evolving series like the consumer price indexes show smaller prediction errors for BSM's as compared to series with ups and downs and a pronounced seasonal pattern. For the latter ARIMA models give a smaller sum of squared errors over all prediction horizons.

Table 9.3a:
Post Sample Sum of Squared Prediction Errors from Conditional Forecasts of Multiplicative Seasonal ARIMA Models, Estimated from 1980 to 1989*

	forecast horizon						
series	90Q1/M3	90Q2/M6	90Q3/M9	90Q4/M12	91Q1/M3	91Q2/M6	91Q3/M9
EMPLC	0.0310	0.0516	0.1574	0.3309	0.4091	0.4833	0.5374
EMPLA	0.0032	0.0048	0.0109	0.0200	0.0402	0.0431	0.0708
CPICQ	0.0281	0.1323	0.1329	0.2759	0.3294	0.4071	0.4938
CPIAQ	0.0159	0.0168	0.0867	0.0889	0.0925	0.0939	0.1005
IPCQ	0.8612	1.2296	2.0440	2.7583	2.7780	4.9158	5.0204
IPAQ	0.0713	0.1286	1.5419	1.6158	4.8418	4.9257	4.9515

* all values have to be multiplied by E-03

Table 9.3b:
Post Sample Sum of Squared Prediction Errors from Conditional Forecasts of Basic Structural Models, Estimated from 1980 to 1989*

	forecast horizon						
Series	90Q1/M3	90Q2/M6	90Q3/M	90Q4/M12	91Q1/M3	91Q2/M6	91Q3/M9
EMPLC	0.0380	0.0785	0.2244	0.3586	0.4406	0.5079	0.5428
EMPLA	0.0283	0.0357	0.0377	0.0547	0.0651	0.0661	0.0971
CPICQ	0.0482	**0.0970**	**0.1076**	**0.1236**	**0.2942**	0.5907	0.6606
CPIAQ	**0.0047**	**0.0061**	**0.0772**	**0.0863**	**0.0864**	**0.0897**	**0.1005**
IPCQ	1.3747	1.4065	3.0233	3.6151	5.0508	6.3356	6.3479
IPAQ	0.4266	0.4486	2.2576	2.2577	5.1685	5.9971	6.2658

* all values have to be multiplied by E-03

The results from generalized Chow tests are presented in tables 9.4a and 9.4b. They do not indicate any data mining for ARIMA models. Significant values can only be found for predictions of Canada's consumer price index from the fourth quarter of 1990 onwards. In this case predictions from both models suffer from an unforeseeable upward movement in the series. The explanation for this level shift can be given by the introduction of a 7 % Federal Sales Tax in January 1991.

Table 9.4a:
Generalized Chow Test of Structural Breaks between Estimation Period 1980–1989 and Prediction Periods for Multiplicative Seasonal ARIMA Models*

	forecast horizon						
series	90Q1/M3	90Q2/M6	90Q3/M9	90Q4/M12	91Q1/M3	91Q2/M6	91Q3/M9
EMPLC	0.33	0.27	0.55	0.87	0.86	0.85	0.81
EMPLA	0.18	0.13	0.20	0.28	0.45	0.40	0.57
CPICQ	1.31	3.10	2.07	3.23+	3.08+	3.18+	3.30+
CPIAQ	0.37	0.20	0.68	0.52	0.44	0.37	0.34
IPCQ	1.21	0.86	0.95	0.96	0.78	1.15	1.00
IPAQ	0.07	0.06	0.48	0.38	0.91	0.77	0.66

* the statistics are distributed F(k, 107) for monthly and F(k,35) for quarterly data, where k indicates the number of forecast steps. For example 90Q1 is distributed F(1,35);

\+ indicates significant value

Table 9.4b:
Generalized Chow Test for Structural Breaks between Estimation Period 1980–1989 and Prediction Periods for Basic Structural Models*

	forecast horizon						
series	90Q1/M3	90Q2/M6	90Q3/M9	90Q4/M12	91Q1/M3	91Q2/M6	91Q3/M9
EMPLC	0.39	0.40	0.76	0.92	0.90	0.86	0.79
EMPLA	1.32	0.84	0.59	0.64	0.61	0.52	0.65
CPICQ	3.16	3.18	2.35	2.03	3.87+	6.49+	6.23+
CPIAQ	0.12	0.09	0.66	0.55	0.44	0.37	0.33
IPCQ	2.06	1.05	1.52	1.36	1.52	1.60	1.37
IPAQ	0.52	0.27	0.92	0.69	1.27	1.23	1.10

* the statistics are distributed F(k, 107) for monthly and F(k,35) for quarterly data, where k indicates the number of forecast steps. For example 90Q1 is distributed F(1,35);

\+ indicates significant value

A further post sample predictive test can be carried out by extrapolating the series multiple steps ahead, starting with the first quarter of 1990. The forecast horizon has been restricted to two years. This interval seems to be an appropriate horizon in terms of short term or demand related economic decisions. The sum of squared errors for several forecasting horizons from these so called unconditional forecasts are given in tables 9.5a and 9.5b. With the exception of the Austrian consumer price index, ARIMA models give a smaller value for the sum of squared errors. That means, the forecast functions from ARIMA models track the series more adequately relative to those from BSM's. However, figures 9.1 to 9.6 provide also graphs of the unconditional forecasts of the series from 1990 to 1991. Evidently both methods give similar forecast functions for all the series. For example the forecast functions of the Canadian employment series in figure 9.1 are almost identical.

Table 9.5a:
Post Sample Sum of Squared Prediction Errors from Unconditional Forecasts of Multiplicative Seasonal ARIMA Models, Estimated from 1980 to 1989*

	forecast horizon						
series	90Q1/M3	90Q2/M6	90Q3/M9	90Q4/M12	91Q1/M3	91Q2/M6	91Q3/M9
EMPLC	0.0232	0.2535	1.1058	4.1331	10.4268	18.5591	28.3317
EMPLA	0.0124	0.0267	0.0655	0.2114	0.3963	0.5945	0.9040
CPICQ	0.0281	0.0365	0.0472	0.0602	0.5076	0.6816	0.6975
CPIAQ	0.0159	0.0251	0.1551	0.2535	0.4577	0.6788	0.9853
IPCQ	0.8612	1.2068	3.9629	13.3942	24.0668	30.0132	35.0497
IPAQ	0.0713	0.3278	0.7934	0.9619	5.1127	9.8441	19.3892

* all values have to be multiplied by E-03

Table 9.5b:
Post Sample Sum of Squared Prediction Errors from Unconditional Forecasts of Basic Structural Models, Estimated from 1980 to 1989*

	forecast horizon						
series	90Q1/M3	90Q2/M6	90Q3/M9	90Q4/M12	91Q1/M3	91Q2/M6	91Q3/M9
EMPLC	0.0306	0.3324	1.4234	4.7093	11.3166	20.1768	31.0742
EMPLA	0.1047	0.2140	0.3567	0.6265	1.2574	1.9338	2.7662
CPICQ	0.0488	0.0789	**0.1566**	0.3879	1.5923	2.7108	3.4973
CPIAQ	**0.0048**	**0.0063**	**0.1092**	**0.1727**	**0.2830**	**0.4053**	**0.6084**
IPCQ	1.3744	2.3651	7.5042	16.7120	35.6299	45.8771	56.1566
IPAQ	0.4256	0.5437	1.2413	1.4008	6.3145	14.7402	29.2338

* all values have to be multiplied by E-03

Up to now, the results suggest to prefer ARIMA models over BSM's at least concerning series with ups and downs and a pronounced seasonal pattern. But the difference in the forecast performance is small and the question arises, whether it is important or not. An answer can be given by means of a forecast encompassing test described in Chong and Hendry [1986]. The basic idea of this test is simple. If forecasts from one model cannot be improved by forecasts from another model, the latter contains no additional information. It is said to be forecast encompassed by the first. Or the other way round, if one model is not able to forecast encompass a rival model, the latter contains some information that is not incorporated into the former. Concerning a time series approach the information with regard to exogenous variables is identical for both models. They differ just in their underlying structure. If an ARIMA model fails to forecast encompass a BSM, there is some useful information contained in the structure of the BSM to improve forecasts from the ARIMA model. This reasoning gives a straightforward test regression of the following kind

$$y_t = \alpha_1 \hat{y}_t + \alpha_2 \tilde{y}_t + e_{1t}, \tag{9.14}$$

where $\hat{y}_t$ are one-step-ahead predictions from the ARIMA model, $\tilde{y}_t$ are one-step-ahead predictions from the BSM and e_{1t} is the regression error. In this case the forecast encompassing test for the ARIMA model versus the BSM model is an F-test for $\alpha_1 = 1, \alpha_2 = 0$. Under the null, ARIMA predictions forecast encompass BSM predictions, i.e. $\alpha_1 = 1$ and $\alpha_2 = 0$ and the regression can be simplified to a t-test on $\alpha_2 = 0$ in a regression of prediction errors from an ARIMA model on predictions from the BSM

$$y_t - \hat{y}_t = \hat{e}_t = \alpha_2 \tilde{y}_t + e_{2t}. \tag{9.15}$$

In small samples the result of such a test is almost certainly affected by the variances of the estimated parameters from the original models. However, the estimated parameters, the corresponding t-statistics, Durbin-Watson statistics and Box-Ljung Q-statistics from regression (4.2) are given in table 9.6a. Autocorrelation in residuals would lead to an overestimated t-statistic, this problem arises in several equations. Nevertheless, ARIMA models forecast encompass predictions from BSM's at the 5% level. The only exception is the Canadian employment series, but evidence there is very weak.

Because the difference between sums of squared prediction errors from both models is not very large, we reversed the null hypothesis and run a set of regressions of type (4.2), where the null was $\alpha_2 = 1$ and $\alpha_1 = 0$. Thus we tested whether the forecasts from BSM's encompass ARIMA models and regressed prediction errors from BSM's on predictions from ARIMA models. The results in table 9.6b show, that the BSM's also forecast encompass their related ARIMA models.

Table 9.6a:
Parameter Estimates and t-values from a Regression of ARIMA One-step-ahead Prediction Errors on BSM One-step-ahead Predictions

series	α_2	t-statistic*	DW-statistic	Q-statistic
EMPLC	-.00013	-2.10	2.03	.08
EMPLA	.00001	.57	1.39	.45
CPICQ	-.00009	-.13	1.56	.16
CPIAQ	.00031	1.04	2.75	.27
IPCQ	-.00045	-.19	1.81	.79
IPAQ	-.00246	-1.24	2.73	.47

* 5 % crit.val. for 20 degrees of freedom (monthly series) is +/- 2.09
5 % crit.val. for 6 degrees (quarterly series) +/- 2.45

Table 9.6b:
Parameter Estimates and t-values from a Regression of BSM One-step-ahead Prediction Errors on ARIMA One-step-ahead Predictions

series	α_2	t-statistic*	DW-statistic	Q-statistic
EMPLC	-.00012	-1.87	2.41	.55
EMPLA	.00002	.74	1.17	.86
CPICQ	-.00016	-.20	2.10	.68
CPIAQ	.00016	.52	2.90	.41
IPCQ	-.00293	-1.24	2.14	.56
IPAQ	-.36993	-1.84	1.91	.92

* 5 % crit.val. for 20 degrees of freedom (monthly series) is +/- 2.09
5 % crit.val. for 6 degrees (quarterly series) +/- 2.45

4 Conclusions

The possibility of frequent structural changes in relations among economic variables during the transition period in Eastern Europe suggests that the use of univariate forecasting methods may be more appropriate both for forecasting purposes and as a convenient data-description tool in support of a decision-making process. Furthermore, during the early stages of economic transition an additional problem may arise in that only short and seasonally unadjusted data are likely to be available. Forecasting such time series is always a difficult task, because statistical methods rely on large-number theory and seasonality introduces further problems. Structural time series models can be viewed as multiple regression models with explanatory variables like of a time trend and a set of seasonal dummies, but with parameters that are time-variant. This gives a suitable framework for statistical model checking procedures as compared to reduced form ARIMA models. Further, the components of the Basic Structural Model can be interpreted in an economically sensible way. Such an interpretation can help to reduce some uncertainty in economic decision making in Eastern Europe. Attributing a shock to a change in the level, slope, or seasonal component of a variable should help the decision maker to get more information about the actual state of the economy and to develop an adequate response scheme.

In this paper we provide some new evidence about the forecasting performance of the structural time series model, with a special focus on short and seasonally unadjusted time series. The results show for all but one of the monthly series smaller in-sample prediction error variance of Basic Structural Models as compared to ARIMA models. These results suggest that the structural model captures losses in information within sample in a better way than the ARIMA model. The relatively better in-sample fit of the Basic Structural Model contrasts with the better out-of-sample performance of ARIMA models. Squared errors from one-step-ahead predictions are smaller for ARIMA models, except for the smoothly evolving consumer price series. These results differ from those in Harvey and Todd [1983], but are generally in conformity with Nerlove et al. ([1979], p. 284ff.) and Garcia-Ferrer and Del Hoyo [1992]. But as pointed out by Nerlove et al., the difference largely depends on the criteria used for the comparison. Moreover, the forecast functions of both models are very similar and eyeballing suggests that there is no big difference. The lack of clear superiority of one method over another seems to be confirmed by a statistical test on forecast encompassing. According to this test, both models forecast encompass each other at the 5 % significance level, implying that neither forecast can be improved by information contained in the other model. Thus forecasts from the BSM are at least as good as the ones from ARIMA models.

The analysis in this paper can be extended in a number of directions. For example it will be useful to present evidence for a larger set of series than evaluated in this paper. One may consider using the 1001 series that the Makridakis et al. competition utilized to provide a more complete comparison of the forecasting performance, though such a comparison may have a different objective than that pursued in this paper. Another extension will require comparing the performance of the two methods using data from East European economies. Efforts are under way to explore some of these extensions and we hope to report some results in the near future.

Figure 9.1:
Canada Employment

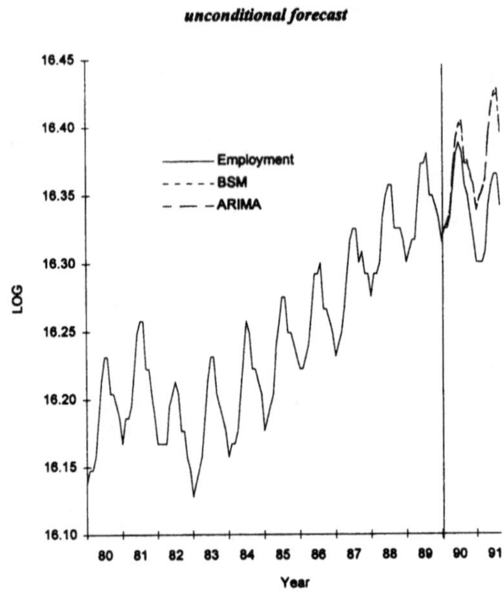

Figure 9.2:
Austria Employment

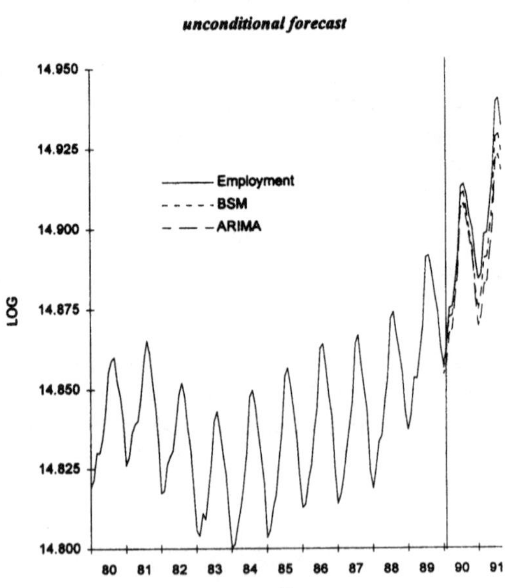

Figure 9.3:
Canada Consumer Price Index
unconditional forecast

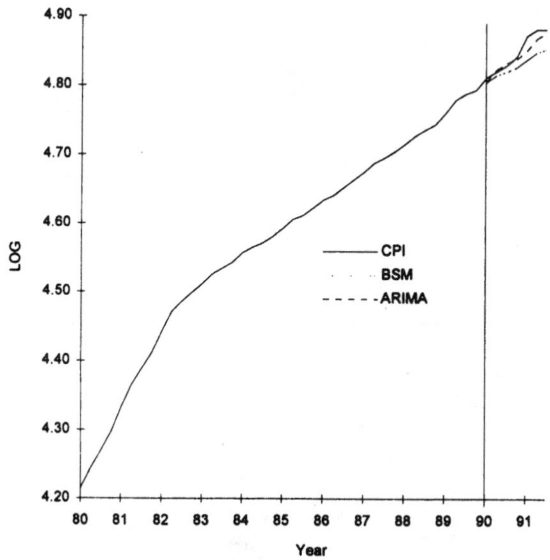

Figure 9.4:
Austria Consumer Price Index
unconditional forecast

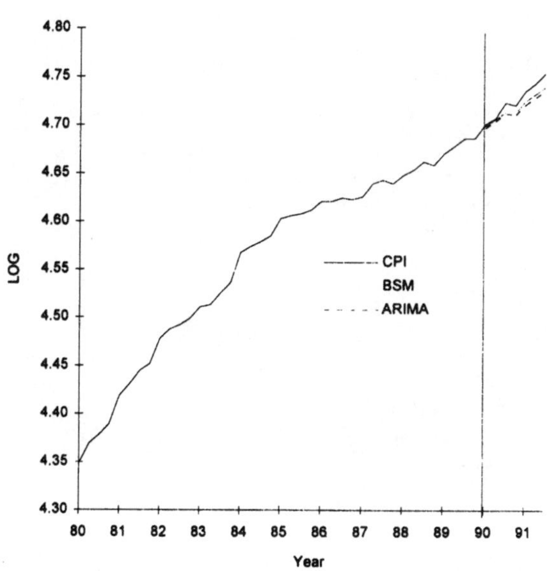

Figure 9.5:
Canada Industrial Production
unconditoinal forecast

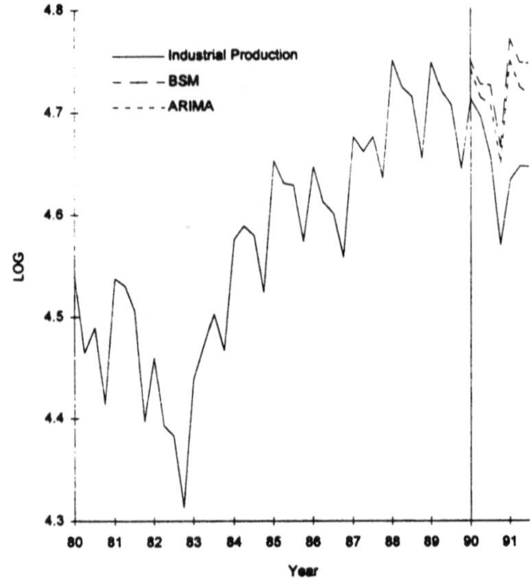

Figure 9.6:
Austria Industrial Production
unconditoinal forecast

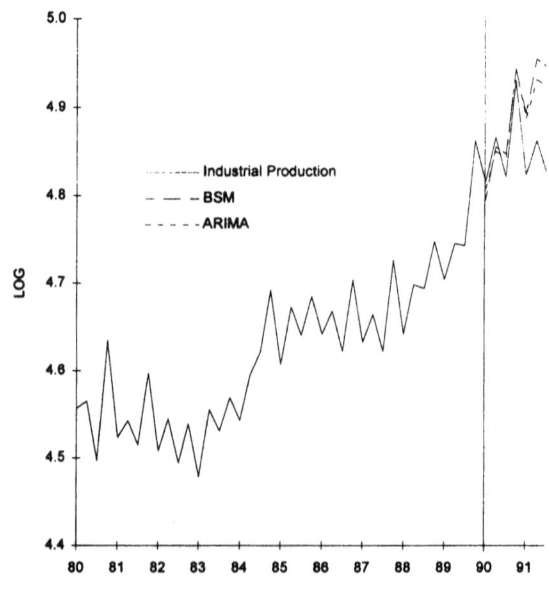

References

Box, G.E. and G.M. Jenkins (1976): "Time Series Analysis Forecasting and Control." Holden Day, Oakland CA.

Box, G.E. and G.C. Tiao (1976): "Comparison of Forecast and Actuality." *Applied Statistics*, **25**: 195–200.

Brockwell, P.J. and R.A. Davis (1991): "Time Series: Theory and Methods." second edition, Springer, New York.

Chong, Y.Y. and D.F. Hendry (1986): "Econometric Evaluation of Linear Macro-Economic Models." *Review of Economic Studies*, **53**: 671–90.

Engle, R.F. (1978): "Estimating structural models of seasonality." In: A. Zellner (ed.), "Seasonal Analysis of Economic Time Series", Bureau of the Census, Washington, D.C., pp. 281–308.

Fildes, R. (1983): "An evaluation of Bayesian forecasting." *Journal of Forecasting*, **2**: 137–150.

Fildes, R. (1984): "Bayesian forecasting." In: S. Makridakis et al. (eds.) "The Forecasting Accuracy of Major Time Series Methods", Wiley, New York, pp. 221–244.

Garcia-Ferrer, A. and J. Del Hoyo (1992): "On Trend Extraction Models: Interpretation, Empirical Evidence and Forecasting Performance." *Journal of Forecasting*, **11**: 645-665.

Harrison, P.J. and C.F. Stevens (1971): "A Bayesian approach to short term forecasting." *Operational Research Quarterly*, **22**: 341–362.

Harrison, P.J. and C.F. Stevens (1976): "Bayesian forecasting." *Journal of the Royal Statistical Society B*, **38**: 205–247.

Harvey, A.C. (1984): "A unified view of statistical forecasting procedures." *Journal of Forecasting*, **3**: 245–275.

Harvey, A.C. (1989): "Forecasting, Structural Time Series Models and the Kalman Filter." Cambridge University Press, Cambridge.

Harvey, A.C. and G.D.A. Phillips (1979): "The estimation of regression models with autoregressive-moving average disturbance." *Biometrika*, **66**: 49–58.

Harvey, A. and N. Shepard (1993): "Structural Time Series Models." to appear in: G.S. Maddala; C.R. Rao and H.D. Vinod (eds.) *Handbook of Statistics*, **11**: Econometrics, North-Holland, Amsterdam.

Harvey, A.C. and P.H.J. Todd (1983): "Forecasting economic time series with structural and Box-Jenkins models: a case study." *Journal of Business and Economic Statistics*, **1**: 299–307.

Lewandowski, R. (1984): "Lewandowski's FORSYS Method." In: S. Makridakis et al. (Eds.), "The Forecasting Accuracy of Major Time Series Methods," Wiley, New York, pp. 245–254.

Nelson, C.R. (1988): "Spurious Trend and Cycle in the State Space Decomposition of a Time Series with a Unit Root." *Journal of Economic Dynamics and Control*, **12**: 475–488.

Nelson, C.R. and C.I. Plosser (1982): "Trends and Random Walks in Macroeconomic Time Series." *Journal of Monetary Economics*, **10**: 132–162.

Nerlove, M. (1972): "Lags in Economic Behavior." *Econometrica*, **40**: 221–251.

Nerlove, M.; D.M. Grether and J.L. Carvalho (1979): "Analysis of Economic Time Series." Academic Press, New York.

Schweppe, F. (1965): "Evaluation of likelihood functions for Gaussian signals." *IEEE Transactions on Information Theory*, **11**: 61–70.

Watson, M.W. (1986): "Univariate Detrending Methods With Stochastic Trends." *Journal of Monetary Economics*, **18**: 49–75.

Part III

Case Studies

Chapter 10

Mobile Sellers and Oligopoly: An Empirical Analysis of the Foreign Exchange Market in Poland, 1988–1989*

WOJCIECH W. CHAREMZA AND GEORGE NORMAN

1 Introduction

One of the distinctive financial institutions which, in the period of the post-planned economy, has become especially evident in Eastern Europe, is a free and highly decentralized domestic market for foreign currencies. Markets like these have existed in Eastern Europe for a long time, supplying foreign currencies to local buyers, who in turn use them for financing their foreign travel (often for commercial purposes), domestic consumption of goods in short supply, investment accumulation and as a speculative asset. Sellers obtain their supply of foreign currencies from foreign tourists, diplomats and from fellow citizens working abroad (for detailed analyses of the currency markets in Eastern Europe see e.g. Charemza and Ghatak [1990], Cowitt [1991], Vanous (forthcoming) and Starzec [1983]). Of the many such East European markets

*We wish to thank Derek Deadman and Klaus Ritzberger, Invited Discussant of our paper, for their helpful comments and suggestions. They are not responsible for any remaining deficiencies. W.W. Charemza gratefully acknowledges financial support of ESRC Grant No. R000 23 1827.

those in Poland are the most highly developed, where finally in March 1989 they were legalized.

The purpose of this study is to develop a theoretical and empirical model of price setting in these markets. Each national market consists of a series of local markets, each local market being located in a major conurbation. The individual local markets may be interconnected to varying degrees. More specifically, it is conjectured that the level of prices for foreign exchange in a particular local market can be affected by mobile sellers from other markets, who can enter the local market if the expected returns from entering the market exceed the cost of entry. The analysis is structured as follows. Section 2 presents the formal economic model underlying subsequent analysis. In section 3 an econometric model of foreign exchange prices is developed on the basis of the theoretical findings derived in section 2. Concepts of unrelated, dominant, homogeneous in aggregate, and broadly and narrowly symmetric markets are formulated as testable hypotheses. These hypotheses are subsequently tested in section 4, where an econometric model of price setting is formulated and estimated for five local markets in Poland, using weekly data from October 1988 to October 1989. Our main conclusions are summarized in section 5.

2 Sellers, Mobility and Entry Costs

Since the foreign exchange market in any one country in Eastern Europe is characterized by a number of sellers who are active at different locations, explicit attention should be paid to the oligopolistic and spatial characteristics of the market. Buyers of foreign currencies, usually households and small private entrepreneurs, are assumed to be spatially distributed. For convenience they can be thought of as being concentrated in the major cities, each of these cities being regarded as an individual (or local) market. Since the majority of buyers are small purchasers, travel costs prevent them from travelling to another conurbation in search of a cheaper price. In other words, they are likely to buy foreign currency locally.

Sellers are also spatially distributed, but can be expected to be more mobile than buyers. They are assumed to be limited in number, with each seller expecting to supply a number of buyers. The currency supplied in any local market (conurbation) is assumed to be homogeneous, although in practice various Western and Eastern currencies are traded.

Our focus in subsequent analysis will be on interconnections between local markets. Thus whenever we refer below to a "market" or to "market i" we shall mean a local market, i.e. a market in a particular conurbation. Competition between sellers in any particular market is modelled as a Cournot game. Each seller sets the quantity to be supplied to the market that will maximize his/her profit, given that all the other sellers to the market are acting in a similar, profit-maximizing, quantity-setting manner[1].

Additionally, our model is based upon the following assumptions:

Assumption 1: All buyers in all markets are identical, with individual inverse demand function for foreign currency of the form:

$$p_i = g(Q_i); \qquad g' < 0 \tag{10.1}$$

where p_i denotes a sellers' price on the i-th (local) market and Q_i is the corresponding aggregate quantity supplied. This simplifying assumption means that any price disparities across the various local markets are not merely the consequence of differences in consumer tastes.

Assumption 2: There are D_i buyers in market i and these buyers purchase only from sellers operating on the i-th market.

Assumption 3: Sellers incur constant marginal costs and may also incur fixed costs of entry to market i. The costs for a seller s to market i are:

$$C_i(s) = F_i(s) + m_i(s)q_i(s), \tag{10.2}$$

where

$$\begin{aligned} F_i(s) &= \text{fixed costs of entry of seller } s \text{ to market } i. \\ q_i(s) &= \text{quantity sold by seller } s \text{ in market } i. \\ m_i(s) &= \text{marginal costs of seller } s \text{ in market } i. \end{aligned}$$

i.e. costs of supplying market i are assumed to be unaffected by costs of supply to another market j. The spatial nature of the aggregate national market lies in the entry costs $F_i(s)$. It is worth noting that this treatment is rather different from the conventional spatial model in which costs of transporting the commodity to a market are typically related not just to distance but also to the quantity transported.

---

[1]In the absence of institutional/empirical detail there are no compelling theoretical reasons why competition between oligopolistic sellers should be based upon quantity rather than price; see the debate on the rationale for Cournot and Bertrand competition (e.g. Daughety [1985]). Recent analysis suggests, however, that if price-competing sellers must pre-commit capacity, the outcome of such capacity-price competition will approach the Cournot equilibrium; see Kreps and Scheinkman [1983] and Vives [1986].

Assumption 4: There are N_i sellers who choose to supply market i.

For a seller s in S_i, where S_i is the set of sellers on market i, the profit function is given by:

$$\pi_i(s) = (g(Q_i) - m_i(s))q_i(s) - F_i(s), \tag{10.3}$$

where $Q_i = \sum_{s \text{ in } S_i} q_i(s)$.

With the assumptions (1) – (4), standard analysis (see e.g. Henderson and Quandt [1980], pp. 202–205) gives the Cournot equilibrium condition in market i from the pricing equation:

$$P_i(p_i^*, N_i) = p_i^* \left[1 - \frac{1}{N_i|\eta_i(p_i^*)|}\right] - \overline{m}_i = 0, \tag{10.4}$$

where:

$|\eta_i(p_i)|$ = absolute value of price elasticity of individual demand in market i at price p_i,

$\overline{m}_i$ = $\sum_{s \text{ in } S_i} m_i(s)/N_i$ = 'average' marginal costs of sellers in i

p_i^* = price in market i at the Cournot equilibrium.

This pricing equation can be rewritten as:

$$p_i^* = \left[\frac{N_i|\eta_i(p_i^*)|}{N_i|\eta_i(p_i^*)| - 1}\right] \overline{m}_i, \tag{10.4'}$$

to indicate the familiar result that price in market i is a form of 'mark-up' on the average marginal costs of the sellers in market i.

Output by seller s in market i is then:

$$q_i^* = \frac{[p_i^* - m_i(s)]}{|g_i'|} \text{ for } s \text{ in } S_i, \tag{10.5}$$

where $g' = \frac{dg}{dQ_i}$,

and profit gross of entry costs is:

$$\pi_i^* = \frac{D_i\,[p_i^* - m_i(s)]^2}{|g'i|} \text{ for } s \text{ in } S_i. \tag{10.6}$$

A necessary condition for seller s to sell in market i is that gross profit to that seller from market i at least covers the entry costs. From equation (10.6) this requires:

$$s \text{ in } S_i \Rightarrow [p_i^* - m_i(s)]^2 \geq \frac{F_i(s)|g_i'|}{D_i}. \tag{10.7}$$

The impact on equilibrium price of an increase in the number of sellers in market i can be derived from equation (10.4):

$$\frac{dp_i^*}{dN_i} = \frac{\delta P_i/\delta N_i}{\delta P_i/\delta p_i^*}, \tag{10.8}$$

(In deriving equation (10.8) we follow convention and treat N_i as a continuous variable.)

Given that the inverse demand function is 'well behaved', i.e. that $\frac{d_i|\eta_i|}{dp_i} \geq 0$, then additional sellers in market i will cause the price in market i to fall if:

$$\frac{dp_i^*}{dN_i} < 0 \quad \text{if} \quad \frac{p_i^*}{N_i^2|\eta_i|} > \frac{d\overline{m}_i}{dN_i}. \tag{10.9}$$

Condition (10.9) is satisfied if the additional sellers' marginal costs are lower than the marginal costs of already operating sellers, i.e. if $d\overline{m}_i/dN_i < 0$. Further, Frank [1965] shows that condition (10.9) holds even if entrants have higher costs than the existing sellers, provided only that equation (10.7) is satisfied for each entrant. Examination of the equilibrium conditions (10.4'), (10.6) and (10.7) indicates that there is unlikely to be uniformity in prices across the various (local) markets. Such uniformity will arise only if all sellers sell in all markets, in which case the problem reduces to the trivial case of a single, unified (national) market. The more likely situation is one in which different markets contain different populations of sellers. To see why this is so, it is convenient to think of the population of sellers S_i to a particular market i as consisting of two sets: S_i^i of indigenous sellers and S_i^f of outside sellers, where one might assume:

$$\begin{aligned} F_i(s) &= 0 \quad \text{for } s \text{ in } S_i^i \\ F_i(r) &> 0 \quad \text{for } r \text{ in } S_i^f \end{aligned} \tag{10.10}$$

For the outside sellers to choose to supply market i, condition (10.7) must be satisfied. This is more likely to be the case where travel to market i is 'easy' and market i is 'large'.

Consider, for example, a simple case in which there are two markets i and j and two sellers s and r, who are indigenous to markets i and j respectively. Assume further that the following conditions hold:

$$p_j^* - m_j(s) > 0; \left[p_j^* - m_j(s)\right]^2 \geq \frac{F_j(s)|g_j'|}{D_j} \tag{10.11}$$

$$p_i^* - m_i(s) > 0; [p_i^* - m_i(s)]^2 \geq \frac{F_i(s)|g_i'|}{D_i}$$

Then both sellers are able to sell in market j, but seller r will not contemplate selling in market i. Such an outcome is more likely to arise the larger is D_j relative to D_i and the greater are the costs of access to market i.

Our analysis suggests, therefore, that there need not be symmetry in price equilibria across all markets, and there there need not be symmetry in price interactions across markets. The latter point is of particular importance when we turn to empirical analysis. We can envisage three broad cases of price interaction between individual markets:

(i) Broad symmetry: Markets are broadly similar in accessibility, so that outside sellers will enter any of the relevant markets in response to the possibility of profits. Entry to market i by sellers who could also supply market j is more likely the higher is the price differential in favor of market i, but entry will, as we have seen, depress this price differential.

(ii) One-way intervention: Sellers will contemplate entering market j from market i, but not vice versa; this is just the case illustrated by equation (10.11).

(iii) Broad independence: This could arise for one of two reasons. First, price in market i may be broadly independent of prices elsewhere because market i is a dominant market, in which all sellers can consider selling. Technically, condition (10.7) is satisfied for most of the sellers. Hence, few new sellers are likely to enter the dominant market and their impact on price will be small. Alternatively, price in market i may be broadly independent of prices elsewhere if market i is small and relatively inaccessible. In that case costs of entry will be relatively high and condition (10.7) is unlikely to be satisfied for outside sellers.

3 Econometric Implications

The theoretical model described in section 2 provides scope for an empirical analysis of interactions between local foreign exchange markets. As is argued in section 1, such markets exist in virtually all Eastern European countries, but at present they are legalized only in Poland, for which some reliable data is available. Bearing in mind that this data relates to prices and not quantities, profits, costs, and number of sellers, some additional assumptions have to be made. In particular, we assume that the number of sellers operating in each market is proportional to the population size of the particular conurbation. We also assume that relations above (equations (10.4) – (10.7)), describing the price-cost margin for a prospective outside entrant can be approximated by the difference between logarithms of the local and 'outside' price, i.e. price from the market in which the outside sellers normally operate.

Let p_{it}, p_{jt} denote logarithms of foreign exchange prices on markets i and j respectively in time t. Suppose that for all local markets price series are integrated of order 1, i.e. are stationary in first differences. If market sellers can use price information from other markets with a one-period delay, and series p_{it}, p_{jt} are cointegrated of order (1,1) with cointegrating vector (1, -1), i.e. if their difference is stationary, the following simple two-equation model can be formulated for each pair of local markets:

$$\Delta p_{it} = \alpha_{ij}\Delta \overline{p}_{it} + \beta_{ij}(\overline{p}_{it-1} - \overline{p}_{jt-1}) + \varepsilon_{ijt}, \quad (10.12)$$

$$\Delta p_{jt} = \alpha_{ji}\Delta \overline{p}_{jt} + \beta_{ji}(\overline{p}_{jt-1} - \overline{p}_{it-1}) + \varepsilon_{jit}, \quad (10.13)$$

where Δ is the first difference operator, ε_{ijt}, ε_{jit}, are error terms and $\overline{p}_{it}(\overline{p}_{it})$ stands for an aggregate price index formulated for all the local markets exclusive of the i-th (j-th) market. The price indices are computed with the use of population sizes of particular conurbations as weights. Variable $\overline{p}_{it}$ is defined as:

$$\overline{p}_{it} = \sum_{s \text{ in } S_i} \omega_{is} \cdot p_{st}, \quad (10.14)$$

where S_i denotes the set of sellers on all markets different from i, and ω_{is} is a population weight, summing up to unity for each i. If the integration and cointegration conditions hold, all the transformed variables in (10.12) and (10.13) are stationary and error terms are also stationary.

The left hand side and first right hand side terms in each equation of the model (10.12) – (10.13) describe the relationship between price changes on a particular market and price changes in the aggregate of the remaining markets. The third term is a direct consequence of the relations (10.4), (10.5), and (10.6); it indicates an expected change in price in time t on market i (j) as the result of the reaction of mobile sellers to price differences that may occur between markets i and j in time $t-1$.

Expected signs of α_{ij}, α_{ji} are positive and of β_{ij}, β_{ji} negative (see equation (10.9)). Since the aggregate variable $\overline{p}_{it}$ contains p_{jt} and p_{jt} contains p_{it}, equations (10.12) and (10.13) are in general not independent even if the error terms ε_{ijt}, ε_{jit} are independent. This can be shown by e.g. multiplying (10.14) by α_{ij} and dissaggregating (zero covariance between microparameters is assumed):

$$\alpha_{ij}\overline{p}_{-it} = \sum_{s \text{ in } S_{ij}} \overline{\alpha}_{is}\omega_{is}p_{st} + \overline{\alpha}_{ij}\omega_{ij}p_{jt}, \tag{10.15}$$

where S_{ij} is the set of sellers on all markets exclusive of i and j and $\overline{\alpha}_{is}$, $\overline{\alpha}_{ij}$ are microparameters for α_{ij}, since $\alpha_{ij} = \sum_{s \text{ in } S_i} \overline{\alpha}_{is}$. Market i (j) is said to be *homogeneous in aggregate* if the coefficient α_{ij}, (α_{ji}) is equal to unity, i.e. if there is a perfect elasticity of the local market with respect to the aggregate market.

Restrictions imposed on parameters and covariance of the error terms in the above equations give the following interpretation of interrelations between particular markets:

(i) If the covariance between error terms ε_{ijt}, ε_{jit}, denoted by σ_{ij} is equal to zero, markets i and j are said to be *weakly unrelated. According to econometric terminology, where the parameters of interest are those of equation (10.12) and* $\sigma_{ij} = 0$, equation (10.12) constitutes a conditional model for p_{it}, i.e. efficient inference for p_{it} can be made independently from an analogous inference on p_{jt}. This bears some resemblance to the concept of *weak exogeneity* (see Engle, Hendry and Richard [1983]). However, in equations (10.12) and (10.13) the variables p_{it} and p_{jt} are not weakly exogenous for each other, since they enter explicitly in each others equations through $\overline{p}_{it}$ and $\overline{p}_{jt}$. Hence, p_{jt} is weakly exogenous for p_{it} in the Engle, Hendry and Richard sense only if $\overline{\alpha}_{ij} = \sigma_{ij} = 0$.
(ii) If in equation (10.12) $\beta_{ij} = 0$, there is *no influence* from market j to market i and analogously, if $\beta_{ji} = 0$, there is no influence from market i to market j. This corresponds to the Granger concept of lack of non-instantaneous causality (see Harvey [1985], p. 302). Note that p_{jt} instantaneously causes (in the Granger sense) p_{it} unless $\alpha_{ij} = 0$ and *vice versa*, p_{it} instantaneously causes p_{jt}, unless $\alpha_{jt} = 0$. Lack of influence from market j to i means that sellers from market j are not able to affect price on market i. They may, or may not, enter the market i, but their role on that market is marginal. If $\beta_{ij} \neq 0$ and $\beta_{ji} \neq 0$, there is a *broad symmetry* between the both markets, as defined in section (2). If sellers from both markets can affect each others price by the same margin, i.e. if $\beta_{ij} = \beta_{ji}$, there is a *narrow symmetry* between these markets.

(iii) If $\sigma_{ij} = \beta_{ij} = 0$, i.e. if markets i and j are weakly unrelated *and* p_{jt} does not influence p_{it}, market i is said to be *strongly unrelated* with respect to market j. This unrelation can be symmetric, if simultaneously $\beta_{ji} = 0$. If, however, $\sigma_{ij} = \beta_{ij} = 0$ and $\beta_{ji} < 0$, then market i *strongly dominates* over market j; its sellers may affect prices on market j, but sellers from j cannot affect prices on i. The concept of strong unrelation is similar to the strong exogeneity of Engle, Hendry and Richard. Nevertheless, a variable, say, p_{jt} is not strongly exogenous for p_{it} even if it is not strongly unrelated to p_{it}, unless $\overline{\alpha}_{ij} = 0$. Analogously, if $\beta_{ji} < 0$, $\beta_{ij} = 0$, but $\sigma_{ij} \neq 0$, market i *weakly dominates* over market j.
(iv) Markets i and j are said to be *symmetric in aggregate* if $\alpha_{ij} = \alpha_{ji}$.

4 Empirical Analysis: Five Local Markets in Poland, 1988–1989

Model (10.12) - (10.13) has been estimated by the use of 49 weekly data observations from October 1988 to October 1989 for five local foreign exchange markets in Poland, each of different size, importance and of different accessibility[2]. We have analysed markets in the following conurbations (the size of a particular conurbation in thousands of inhabitants is given in brackets together with an abbreviation for its name used hereafter): Czestochowa (253, CZE), Katowice (369, KAT), Kraków (745, KRA), Poznan (586, POZ) and Warsaw (1,671, WAR). As can be seen, markets are of various sizes, with the Warsaw market twice as large as the second biggest one, Kraków. The matrix of approximate road distances between the markets is given in table 10.1. The matrix indicates that three of these conurbations, Czestochowa, Katowice and Kraków are clustered together within a short travel distance, while two remaining markets, Poznan and Warsaw, are relatively distant from the other markets.

The order of integration for particular series has been tested with the use of the Augmented Dickey-Fuller (ADF) test for unit roots. Results are given in table 10.2. In table 10.3 the ADF statistics for testing the stationarity of differences of particular pairs of price series, i.e. cointegration of the series p_{it}, p_{jt}, with the cointegrating vector (1, -1) are presented (since the cointegrating vector is not estimated, the same critical values as before can be used). The

[2]Most of the data were published in weekly 'Gazeta Bankowa'. Some gaps in the series were filled through direct consultations with local dealers. We use price data published for an unorganized, but tolerated, market of individual dealers rather than for private licensed exchange offices ('kantory'). Original data were adjusted in time to allow for a possible cyclical analysis, so that there are always four 'weekly' data in a month. The full set of data is available from us on request

results show that all the price variables, both individual (p_{it}) and aggregate ($\overline{p}_{it}$) are integrated of order 1, i.e their first differences are stationary. Similarly, differences between particular individual prices are stationary. We may then conclude that all the transformed variables which appear in model (10.12) – (10.13) are stationary.

Table 10.1:
Approximate Road Distances between Local Markets (in Kilometers)

Market	CZE	KAT	KRA	POZ	WAR
CZE	0	65	100	280	215
KAT	–	0	80	320	280
KRA	–	–	0	380	295
POZ	–	–	–	0	300

Table 10.2:
Augmented Dickey-Fuller Test for Level of Integration

Market	Levels		First Differences	
(i)	p_{it}	$\overline{p}_{it}$	p_{it}	$\overline{p}_{it}$
CZE	-0.31	-0.57	-3.16*	-3.02*
KAT	-0.60	-0.55	-4.34*	-2.96*
KRA	-0.53	-0.54	-3.21*	-3.05*
POZ	-0.52	-0.56	-3.95*	-3.05*
WAR	-0.52	-0.47	-3.41*	-3.31*

Legend: * denotes values significant at 5 % level of significance

Table 10.3:
Augmented Dickey-Fuller Test for Stationarity of $p_{it} - p_{jt}$

Market	CZE	KAT	KRA	POZ	WAR
CZE	N/A	-6.06*	-3.17*	-5.81*	-3.82*
KAT	–	N/A	-2.99*	-6.33*	-4.82*
KRA	–	–	N/A	-4.39*	-4.82*
POZ	–	–	–	N/A	-3.69*

Legend: See table 10.2.

Table 10.4:

Testing Weak Unrelation between Pairs of Markets ($\sigma_{ij} = 0$)

Market	CZE	KAT	KRA	POZ	WAR
CZE	N/A	2.90*	4.34*	4.14*	0.01
KAT	–	N/A	3.64*	3.84*	0.42
KRA	–	–	N/A	3.35*	-1.16
POZ	–	–	–	N/A	-0.78

Legend: See table 10.2

Although the estimation of the model (10.12) – (10.13) allows for a non-diagonal covariance matrix of the error terms, testing diagonality of the covariance matrix allows for identification of markets which are not weakly unrelated, i.e. markets for which $\sigma_{ij} \neq 0$. Testing is similar to testing weak exogeneity and consists of estimating equation (10.13) under the hypothesis of zero covariance by ordinary least squares, computing residuals and testing these residuals as the omitted variable in equation (10.12) – see Engle ([1984], p. 815). Since the residuals are the only omitted variable, the conventional t-Student test is equivalent to the Lagrange multiplier test. Results are given in table 10.4. The hypothesis of zero covariance cannot be rejected for the Warsaw market in relation with all the others. Hence, we conjecture that the Warsaw market is weakly unrelated with respect to the remaining four. The other four markets, however, are not weakly unrelated, since the hypothesis of zero covariance is rejected in each case.

Ten models of the type (10.12) – (10.13), for each pair of the five markets, have been estimated by the maximum likelihood method allowing (but not assuming) a non-diagonal covariance matrix of error terms[3]. Estimation results and basic diagnostics are shown in table 10.5. The estimates seem to be statistically sensible and trustworthy. All estimates of α_{ij} and α_{ji} are significantly different from zero, the signs of the estimates of β_{ij} and β_{ji} are, as expected, negative, goodness-of-fit is high and there is no indication of first-order autocorrelation. Results of testing homogeneity in aggregate, i.e. applying the t-Student test for testing the hypothesis that $\alpha_{ij}(\alpha_{ji}) = 1$ reveals that Kraków cannot be regarded as homogeneous in aggregate. For Czestochowa and Warsaw the test is not conclusive, with average t-Student statistics of -1.79 and -1.92 respectively.

[3]For estimation we use the TSP-4.1 econometric computer package. The estimation command is LSQ with option HCOV=RG, i.e estimates of standard errors are heteroscedasticity-consistent.

Most of the markets are broadly symmetric, i.e. both corresponding estimates of β_{ij} and β_{ji} are significant. The exception is the smallest of the markets, Czestochowa, which is broadly symmetric only with respect to Warsaw and, in a single case, Warsaw, which may not be broadly symmetric for Katowice. In fact we may conjecture that Czestochowa weakly dominates over Katowice, Kraków and Poznan, whilst Warsaw may strongly dominate over Katowice, since for Katowice and Warsaw the hypothesis that $\sigma_{ij} = 0$ cannot be rejected. This conjecture is not, however, strong, because the t-Student ratio for the β_{ji} estimate in the Warsaw-Katowice model is on the verge of significance.

For testing narrow symmetry, i.e. for the cross-equation restriction $\beta_{ij} = \beta_{ji}$, the likelihood ratio test of the type $LR = 2^*(log(l_u) - log(l_0))$ has been used, where $log(l_u)$ is the value of the loglikelihood function computed for the unrestricted model and $log(l_0)$ stands for the value of the loglikelihood function for the model estimated with the restriction imposed. Under the null hypothesis of valid restriction, the LR statistic has an asymptotic $\chi^2(1)$ distribution (see e.g. Harvey [1985], p. 163). These statistics are given in table 10.6. They show that the hypothesis of narrow symmetry can be rejected for Czestochowa in relation with Kraków and Poznan, which confirms strongly the previous conjecture about Czestochowa weak dominance. For Katowice-Warsaw, however, the fact that the narrow symmetry hypothesis cannot be rejected undermines the previous conjecture about strong dominance of Warsaw over Katowice.

The hypothesis of symmetry in aggregate, i.e. $\alpha_{ij} = \alpha_{ji}$, has been tested with the use of an analogous likelihood ratio test, by comparing the loglikelihood function computed for the unrestricted and restricted models (see table 10.7). It reveals that there is only one pair of markets which can be sensibly interpreted as being symmetric in aggregate: Czestochowa and Katowice (note that for those markets narrow symmetry also cannot be ruled out). Since the distance between these markets is the smallest (see table 10.1) this may indicate low cross-market entry costs and similar structure of market information. Nevertheless, relative costs of entry are presumably not the same on these markets, since the Czestochowa and Katowice markets behave differently in relation with other markets.

Table 10.5:
Max. Likelihood Estimates of Model (10.12) – (10.13)
for all Pairs of Markets

Pairs of markets	α_{ij}	β_{ij}	SSE_{ij}	R^2_{ij}	DW_{ij}
	α_{ji}	β_{ji}	SSE_{ji}	R^2_{ji}	DW_{ji}
	[-1.67]				
	0.871	-0.196	0.0569	0.993	1.88
CZE	(11.31)	(-1.67)			
KAT					
	[-1.00]				
(188.13)	0.919	-0.526	0.0706	0.992	1.70
	(11.26)	(-3.40)			
	[-1.87]				
	0.862	-0.186	0.0576	0.993	1.98
CZE	(11.63)	(-1.14)			
KRA					
	[-4.97]				
(189.02)	0.671	-0.754	0.0782	0.991	1.86
	(10.13)	(-3.96)			
	[-1.96]				
	0.850	-0.094	0.0586	0.993	2.00
CZE	(11.08)	(-0.88)			
POZ					
	[-0.93]				
(177.12)	1.088	-0.542	0.108		
	(11.49)	(-3.99)		0.988	2.06
	[-1.65]				
	0.874	-0.286	0.0520	0.994	1.82
CZE	(11.45)	(-3.38)			
WAR					
	[-2.17]				
(176.46)	0.820	-0.483	0.0973	0.989	2.22
	(9.87)	(-2.13)			
	[-0.95]				
	0.917	-0.325	0.0799	0.991	1.87
KAT	(10.43)	(-2.41)			
KRA					
	[-4.49]				
(175.34)	0.682	-0.436	0.0936	0.989	2.00
	(9.64)	(-2.70)			

Legend: See over.

Table 10.5: (Continued)

Pairs of markets	α_{ij} α_{ji}	β_{ij} β_{ji}	SSE_{ij} SSE_{ji}	R^2_{ij} R^2_{ji}	DW_{ij} DW_{ji}
	[-1.20]				
	0.888	-0.267	0.0840	0.990	1.93
KAT	(9.56)	(-2.21)			
POZ					
	[1.74]				
(177.19)	1.156	-0.929	0.0796	0.991	1.80
	(12.93)	(-7.64)			
	[-0.99]				
	0.916	-0.475	0.0698	0.992	1.76
KAT	(10.70)	(-4.51)			
WAR					
	[-1.92]				
(168.62)	0.824	-0.448	0.1016	0.988	2.39
	(8.95)	(-1.55)			
	[-5.02]				
	0.638	-0.318	0.0963	0.988	2.21
KRA	(8.86)	(-2.86)			
POZ					
	[0.67]				
(167.08)	1.071	-0.435	0.1240	0.987	2.12
	(10.05)	(-3.52)			
	[-4.15]				
	0.723	-0.515	0.0749	0.991	1.94
KRA	(10.84)	(-5.96)			
WAR					
	[-1.50]				
(172.27)	0.879	-0.464	0.0924	0.990	1.87
	(10.87)	(-2.60)			
	[1.75]				
	1.145	-0.724	0.0824	0.990	1.65
POZ	(13.89)	(-7.35)			
WAR					
	[-2.10]				
(169.21)	0.816	-0.519	0.0862	0.990	2.13
	(9.03)	(-2.96)			

Legend: In the first column values of the loglikelihood function are in brackets. In the second and third columns t-Student ratios are in ordinary brackets and in the second column t-statistics for testing the homogeneity hypothesis (that the regression coefficient is equal to unity) are in square brackets. SSE is the sum of squared residuals, R^2 is the determination coefficient (note the absence of constant terms in the equations) and DW is the Durban-Watson coefficient.

Table 10.6:
LR Statistics for Testing Narrow Symmetry $\beta_{ij} = \beta_{ji}$

Market	CZE	KAT	KRA	POZ	WAR
CZE	N/A	1.79	3.88*	4.22*	1.10
KAT	–	N/A	0.19	5.69*	0.02
KRA	–	–	N/A	0.27	0.14
POZ	–	–	–	N/A	1.70

Legend: See table 10.2

Table 10.7:
LR Statistics for Testing Symmetry in Aggregate $\alpha_{ij} = \alpha_{ji}$

Market	CZE	KAT	KRA	POZ	WAR
CZE	N/A	0.56	10.05*	8.82*	0.34
KAT	–	N/A	10.84*	12.07*	0.91
KRA	–	–	N/A	25.04*	1.85
POZ	–	–	–	N/A	8.42*

Legend: See table 10.2

5 Conclusions

The main theoretical finding of this paper is that in an unorganized foreign exchange market with mobile sellers, the local market entry costs of the mobile sellers may determine the change in price on the local market. This result seems to be especially important since the total costs of 'outsiders' can be greater than the costs of the already established sellers, yet the entry of the outsiders may still cause the price to fall. Intuitively, sellers should have a strong incentive to be mobile and to seek profit even in relatively remote markets. Our empirical analysis confirms this effect for 1988–1989.

The theoretical model of mobile sellers in an unorganized foreign exchange market in section 2 allows for formulation and estimation of a feasible and interpretable econometric model, for which the only data required are prices on particular local markets. As expected, most markets appear to be at least broadly symmetric. Our investigation reveals, however, that interrelations and dependencies in the Polish foreign exchange market are rather tangled. The biggest local market, in Warsaw, can be statistically separated from the other markets and can be treated separately. The remaining markets are interrelated and mainly broadly symmetric, although their role in the aggregate differs significantly. It is interesting that behavior of the smallest of the markets, Czestochowa, bears some resemblance to the behavior of the largest one,

Warsaw. In particular the Czestochowa market shows weak dominance over a number of other local markets. This phenomenon can be interpreted as being the result of substantial differences in entry costs between small and large conurbations.

One of the advantages of the method described above is that it can be easily repeated for other countries and for different sets of conurbations. In particular, for Poland, where since 1989 more detailed data on foreign exchange markets have been available, it will be possible to repeat the investigation for a wider set of markets, which should include at least two additional important markets missing from our research due to the lack of sufficient data, namely Gdask and Szczecin.

References

Charemza, W.W. and S. Ghatak (1990): "Demand for money in a dual- currency, quantity-constrained economy: Hungary and Poland." *The Economic Journal*, **100**: 1159–1172.

Cowitt, P.P. (1991): "1988–1989 World Currency Yearbook." International Currency Analysis, Brooklyn, New York.

Daughety, A.F. (1985): "Reconsidering Cournot: The Cournot Equilibrium is Consistent." *Rand Journal of Economics*, **16**: 368–379.

Engle, R.F. (1984): "Wald, likelihood Ratio and Lagrange Multiplier Tests in Econometrics." In: K.J. Arrow and P.H. Intriligator (eds.), *Handbook of Econometrics*, **2**: 776–826.

Engle, R.F., D.F. Hendry and J.-F. Richard (1983): "Exogeneity." *Econometrica*, **51**: 277–304.

Frank, C.R. (1965): "Entry in a Cournot Market." *Review of Economic Studies*, **32**: 245–250.

Harvey, A.C. (1985): "The Econometric Analysis of Time Series." Phillip Allan, Oxford.

Henderson, J.M. and R.E. Quandt (1980): "Microeconomic Theory: A Mathematical Approach." McGraw-Hill, London.

Kreps, D.M. and J.A. Scheinkman (1983): "Quantity Precommitment and Bertrand Competition Yield Cournot Outcomes." *Bell Journal of Economics*, **14**: 326–337.

Starzec, K. (1983): "L'économie polonaise vue à travers des circuits parallèles." Paris, mimeo.

Vanous, J. (forthcoming): "Private Foreign Exchange Markets in Eastern Europe and the USSR." In: G. Grossman (ed.), Studies in the Second Economy in the Communist Countries, Berkeley University Press, Berkeley.

Vives, X. (1986): "Commitment, Flexibility and Market Outcomes." *International Journal of Industrial Organization*, 4: 217–229.

Chapter 11

Quantitative Modeling in the Presence of Structural Breaks: Assessing Energy Demand and Supply for the Soviet Union up to 1995

FRANZ WIRL

1 Introduction

The following paper tries to assess the developments of energy demand/supply in the Soviet Union after the dramatic political break in the historical evolution and accounting for the emerging market structures. The analysis is entirely set within a partial equilibrium framework so that general equilibrium effects are omitted, despite the importance of energy for the overall economy in particular for an energy exporting country like the Soviet Union. For example, fast deregulation stimulates first conservation thus increases exports; these increase of revenues may raise in turn real income which then might lower conservation. Furthermore, this paper treats the Soviet Union as a geographic entity as defined politically prior to August 1991. Hence, the Baltic states and any further dissolving republic is still considered as a part of this Union.

This paper focuses on discussions and results and minimizes technicalities. The empirical data base and the modeling approach are similar to one used in Wirl [1991b]. Section 2 sketches some properties of the command economy and outlines the difficulties this economic system would face in the future. Section 3 discusses economically efficient versus politically efficient deregulation strategies. Section 4 presents an optimistic view – a discontinuous introduction of markets. Section 5 points at potential political obstacles to such an optimistic development. A summary completes this investigation and an appendix documents the empirical framework and its parameters.

2 Properties of the Command Economy

The command economy has a number of distinctive properties. In particular, the overall lack of incentives leads to a tremendously inefficient economy and low standards of living. Indeed, considering this lack of incentives (often reduced to occasional, sometimes subtle coercion) and with the full knowledge of hindsight, it seems fascinating how long these socialist economies and in particular the Soviet Union survived. Thus, with this perfect knowledge of hindsight it seems necessary to explain the temporary success of the socialist economies rather than their final collapse. The following thesis from Sir Karl Popper ([1963] p. 365) "We are good, perhaps a little too good, but we are also stupid; and it is this mixture of goodness and stupidity which lies at the root of our troubles" may help to explain the success of dictatorship in general; he adds (p. 366) "Admittedly, the great dictators did appeal to all sorts of fears and hopes, to prejudices and envy, and even to hatred. But their main appeal was an appeal to a kind of morality. They had a message; and they demanded sacrifice. It is sad to see how easily an appeal to morality can be misused. But it is simply a fact that the great dictators were always trying to convince their people that they knew the way to a higher morality." In other words, the goodwill of people and the politicians appeal to morale and ideals (to participate in the build up of paradise, the classless society) and the ethics of the bourgeois and peasant society (i.e., to work rather than to steal) explains the 'success' at the beginning; later on coercion, raw material exports (oil, gas) and Western aid prevented presumably an earlier collapse. Levy [1990] provides an economist's account and a critique of the lack of the economic profession to foresee these developments[1]. Indeed, even Hayek and von Mises, 'who carried alone the brunt of the anti-central planning position' did not emphasize the lack of incentives but concentrated on 'the estimation marginal cost prices

[1] He offers a public choice explanation: the socialist economies provide so high rewards to economists – they are responsible for all the plans, e.g. assessment of marginal cost prices in a Lerner-type market socialism. In fact, Hayek saw already this connection but, according to Levy [1990], was too polite to his colleagues and attributed to them at maximum intellectual error but never private interest.

by markets and a planning bureaucracy' and the impossibility to process this information centrally. Of course, this aspect was ultimately verified too as the absurd prices not only for energy but for almost all commodities document. Figure 11.1 confirms this prediction for the energy prices; this figure compares the energy prices in the USSR with those of the Federal Republic of Germany (prior to the unification), which has relatively low energy prices; moreover, this comparison uses for the conversion the favorable ratio of 0.55 roubles per one US dollar, which is by a two-digit number off the 'market' rate.

Figure 11.1:
Comparison of (Industrial) Energy Prices

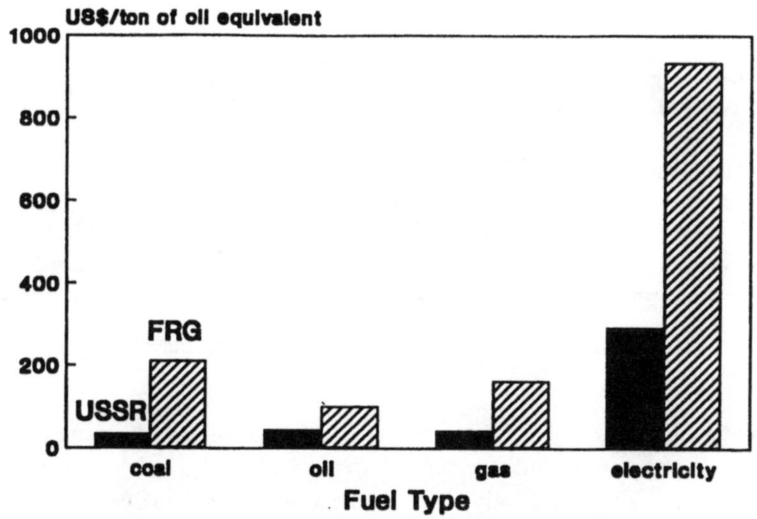

Using the conversion of 0.55 roubles = 1$

Table 11.1 shows some indicators of the Soviet Union and compares them with OECD countries. These numbers indicate low standards of living and 'over-industrialization', more precisely basic industries (e.g., iron and steel) are far more dominant than in any market economy. The assessment of real income, e.g., GDP per capita on the other hand is very difficult. The reasons are the different accounting schemes, net material product (NMP) versus gross domestic product (GDP), the neglect of inflation, the inconvertibility of the rouble, the different quality of products, etc. In fact, GDP assessments range widely, e.g. the CIA suggests an incredibly high GDP (twice that of Portugal) on the upper range while (black) market exchange rates determine the lower bound. One reason for this upward bias (presumably even in the PlanEcon's assessment that will be used in the sequel) seems that they are based on purchasing power parity comparison without discounting sufficiently for the low quality of Eastern European goods.

Table 11.1:
Some Economic Characteristics of the Soviet Union 1987

Steel	per capita	per Net Material Product (NMP)
USSR	572	151
USA	332	31

	Floor Space m^2/capita	Cars per '000	Televisions	Telephones
			persons per receiver	
USSR	15	51	3.2	7.6
USA	50	555	1.2	1.3
Portugal		370	6.2	4.6

GDP per capita ($)	CIA	PLanEcon	Market*)
USSR	8300	3600	225
USA			18300
Portugal			3700

* Based on market exchange rates as reported in Nordhaus [1991].

Energy inefficiency is just a particular example of the overall input inefficiency of the command economy. Figure 11.2 shows the scope for energy conservation. More precisely, this figure shows the relative amount of energy inputs that yields the same output in OECD-Europe. For example, the steel mills in Soviet Union need approximately four times the amount of energy in order to produce a ton of steel compared with an average Western European mill. An exception is private transportation, which was severely rationed in most East European countries. As a consequence, a market economy will here tend to increase rather than to decrease the energy demand. However, rationing seems not that obvious for the Soviet Union since the corresponding intensity (i.e. accounting for the low income) exceeds the one of OECD Europe; the right bar in figure 11.2. However, if we take the U.S. intensity as the denominator to account for the vast size of the Soviet Union, a 'negative' conservation results for road. Furthermore, the shrinking of the energy intensive basic industries offers an important second order effect for aggregate energy conservation.

Figure 11.2:
Energy Efficiencies in OECD-Europe Relative to the USSR and further Potential Reductions; for 'Road', the US Serves as Reference and OECD-Europe Determines 'Potential'

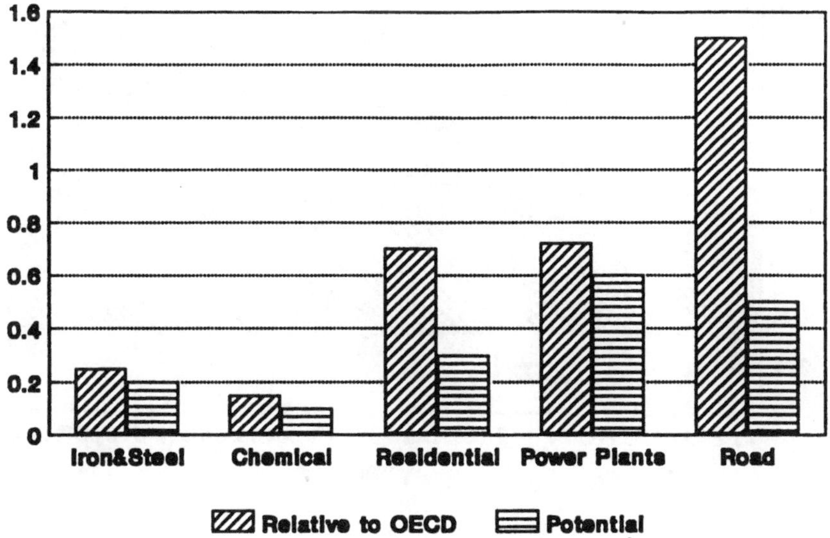

Another remarkable observation of the centrally planned economies is the extraordinary smooth evolution of the historical energy pattern. Thus, USSR's energy demand did not respond to the important events of the global energy markets and in particular, not to the energy price shocks in 1974 and 1979. Indeed, this evolution seems to confirm the allocation rules of a planning economy where firms simply demand more and more energy to fulfill their plan[2]. Figure 11.3 shows the historical evolution and integrates a projection[3] for the hypothesis of a continuation of the command economy of the Brezhnev type. This extrapolation is characterized by moderate but continuous energy demand growth, diminishing contributions of solids and increased market shares for natural gas. Oil consumption grows but only modestly.

[2]This observation applies also at a disaggregated level. For example, the energy efficiency in iron & steel did not improve (in fact slightly deteriorated) over the period 1978–1987.

[3]Technically speaking, the projections for the demand of the command economy are the result of time series analysis, which seems to capture best the tendencies inherent to a command economy; this has been done at a fairly detailed level of twenty-seven flows and eight products. Moreover, published economic indicators even where available are hardly usable, e.g. no inflation is reported officially, outputs are upward biased, and prices do not really matter in a command economy.

Figure 11.3:
Total Primary Energy Requirements of the USSR by Fuel

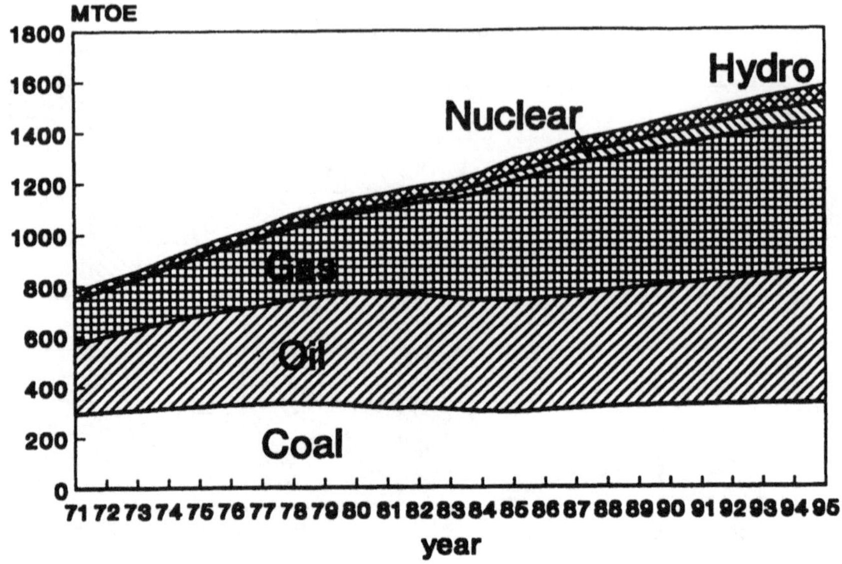

Figure 11.4:
Primary Energy Production of the USSR by Fuel

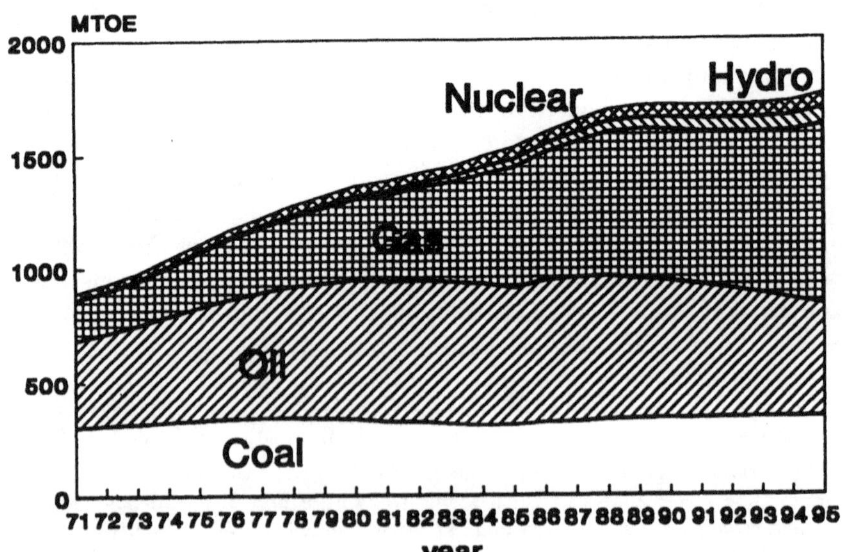

Figure 11.4 is the pendant of figure 11.3 and shows the production of primary energy. These extrapolations reflect a literature survey (see e.g., Stern [1987] and Sinyak [1990]), discussions with various experts (except for crude oil production) and demand constraints. In particular, coal production could increase further but lacks market (domestically and export); similarly, some authors conjecture a higher potential natural gas output, which, however, Western Europe may not be able to absorb.

The hard currency revenues from the oil exports are vital for the Sovietunion, the world's largest oil and energy producer. Figure 11.5 shows the historical production of crude oil, the result from a logistic fit and two other extrapolations. The logistic curve fitted (from 1918-1990) to cumulative production is a phenomenological description of oil production (based on Hubbert [1978]). The empirical finding is reported in figure 11.5, which highlights that this curve yields an excellent fit and seems to cater quite well the recent decline in USSR oil production; extrapolation describes the scenario of an ongoing command economy[4]. Additionally, this figure shows another fairly recent forecast (Sinyak [1990]) and the (endogenous) extrapolation for the hypothesis of a radical transition to a market economy, i.e., when the international oil companies are allowed (and sufficiently rewarded) to clear the present mess. More precisely, it is reasonable to assume that the overall inefficiency of the command economy is also present in the oil industry so that there is a considerable scope for improvement. In fact, considering the low recovery rates from the oil fields, the highly outdated technology, unmotivated engineers and workers, and the conjectured huge resources[5], USSR oil production may recover from the recent sharp declines and may even increase. Of course, this requires the assistance of the international oil companies who in turn will demand well established property rights and guarantees.

[4]This fit overestimates 1990 and seems thus even too optimistic compared with the recent news about the shape of the USSR oil industry.

[5]The estimated parameters of the logistic curve imply an assessment of the ultimate recoverable resource basis of 175 billion barrels (the numerator); this is very conservative compared with a Delphi-poll among experts see Desprairies [1978].

Figure 11.5:
USSR Oil Production – Historical and Forecasts and Estimation for X_t Cumulative Production up to Year t, Sample 1918-1989, Figures in Brackets Denote Absolute t-statistics:

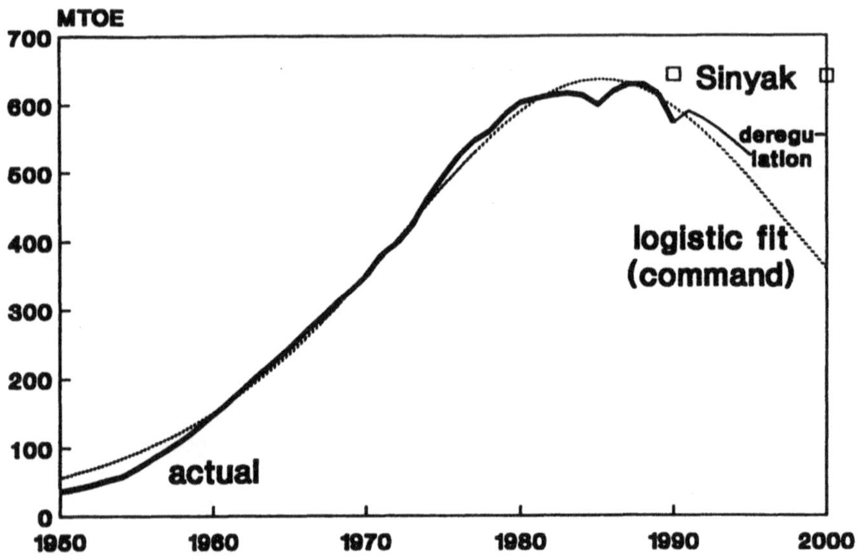

$$X_t = \frac{\underset{(31.7)\ (36.0)\ (78.9)}{173708}}{[1 + 1397^{(-0.107(t-1918))}]} \quad \overline{R}^2 = 0.99$$

The major purpose of this section is to prove that a continuation of the traditional economic system would face extreme difficulties and frictions. Figure 11.6 reports the (hypothetical) energy trade and highlights that the USSR would become a net importer of crude oil. However, financing this amount of crude oil imports seems impossible considering the inconvertibility of the rouble since the gain in gas export revenues cannot offset the loss from the oil revenues let alone, finance the oil imports. Hence, the planning authorities have to expand rationing in order to prevent that the USSR from becoming an oil importer. This in turn affects negatively the output of the oil hungry industry and demotivates people, e.g., because private mobility may become even less attainable for them. Thus, execution of such a scenario is hardly imaginable, even for a Soviet Union of the old regime, let alone in the age of Glasnost.

Figure 11.6:
USSR Net Energy Exports if the Command Economy would Continue

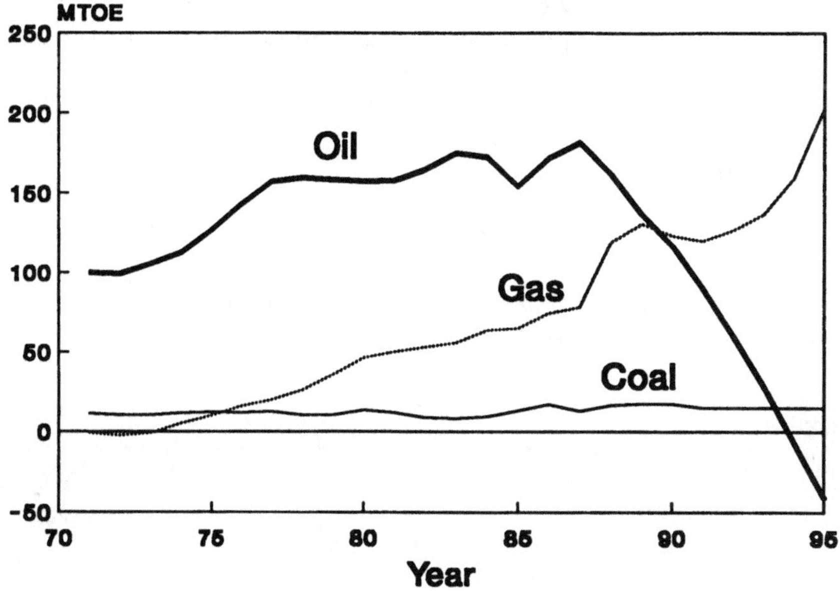

3 Normative and Positive Theory of Deregulation

It is straightforward to prove (at least within a partial equilibrium framework) that there is no justification on grounds of economic efficiency to phase the introduction of market prices. In other words, historical rules and price distortions etc. that influenced past investments (even if they are entirely sunk) of households and firms do not justify a phased transition. And this result – radical deregulations are efficient – holds even for naive, i.e., myopic, consumers.

Figure 11.7 summarizes graphically the result from Wirl [1991a]. The downward sloping curve reflects the equilibrium inverse demand and the upward sloping curve represents the supply, i.e. the marginal cost curve. Now, the efficient government's policy for the hypothesis of perfectly rational consumers is to charge immediately the marginal costs, irrespective of initial demand (which is too high). Thus prices jump immediately such that they equate to marginal costs and then decline along the marginal costs curve until the stationary equilibrium is reached. A sharp initial price increase and then declining prices still holds for myopic consumers but the initial price may be below the full marginal costs. The reason is simple because full marginal costs lead to too excessive conservation investments for consumers that do not foresee the future decline in prices. However, the initial price exceeds in both cases the ultimate equilibrium price, i.e. deregulation must be radical. However, vote maximizing politicians and a myopic electorate that penalizes price increases but appreciates efficiency provides a considerable deviation from the efficient strategy, transiently (starting with a low price) and maybe even persistently (deviations from marginal cost tariffs due populistic motivations, efforts of important pressure groups, etc.). An example of such a political strategy is shown in figure 11.7.

Figure 11.7:
Efficient Deregulation – Rational & Myopic Consumers and Vote Maximizing Deregulation for Myopic Consumers when Adjustment Takes a 'Long' Time

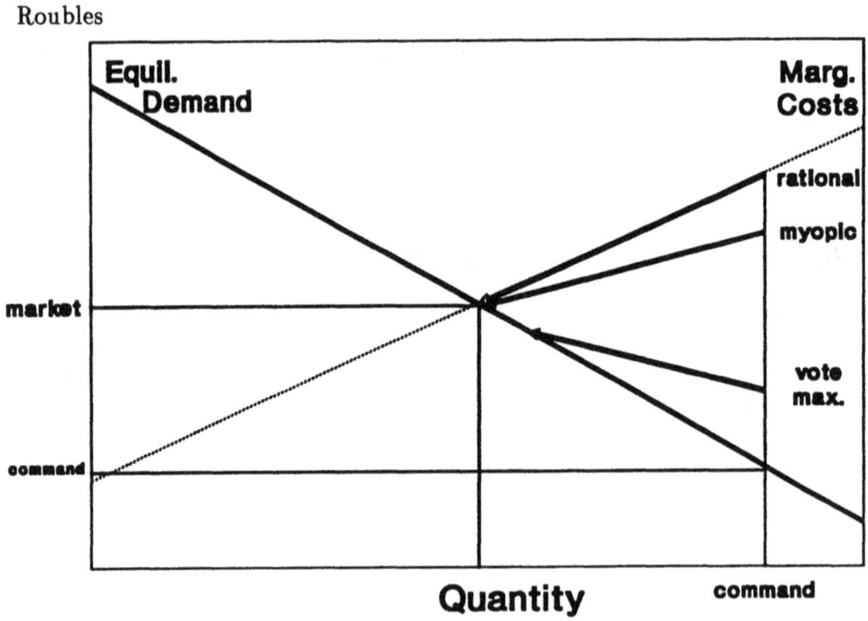

4 An Optimistic View – Radical Introduction of Markets

The following quantification of the USSR's energy demand over the next five years uses the framework outlined in the Appendix. This model is similar to Wirl [1991b] and reflects the following ideas within a standard framework:

> Energy prices influence energy efficiencies and govern fuel substitution, but adjustment is very sluggish. Furthermore, energy efficiency improvements may proceed independent of actual energy prices due to the replacement of capital (the new and now Western technologies will be more efficient) and the overall slack, which is characteristic for a command economy.

The parameters for the scenario of an energy market in the Soviet Union are based on 'experts' opinions, i.e, a literature survey and own judgment rather than on traditional estimation. This is done very carefully and cautiously. This seems the proper way to assess the parameters because several reasons impede an estimation of the energy demand relations from historical data. First, data is either unreliable, not available, or meaningless. However, even if a perfect data set does not justify conventional econometric estimations due to the peculiarities of the command economies and the structural break from a planned to a market economy. Thus, one cannot extrapolate relations that are econometrically fitted over a sample of the now dead command economy[6]. In the absence of estimation, we construct the "energy market" in the Soviet Union from the knowledge obtained from real world energy markets, i.e., from the experience of the industrialized countries, which has been extensively documented in the empirical literature.

The corresponding scenario supposes 'world market prices' for all energy carriers. This seems highly unrealistic even for an optimist. However, we use the favorable exchange rate of 0.55 roubles per US$ to aim for realism. Thus, domestic energy prices move to European price levels on a purchasing power parity basis but do not reflect full opportunity costs. This amounts nevertheless to dramatic real energy price increases of around 300 % and above, see figure 11.1. These price increases are far above those observed within the industrialized countries following the two OPEC oil price shocks. All parameter assumptions are highly conservative such that even this price increase will not move energy intensities fully down to the levels indicated in figure 11.2. Furthermore, economic activity will shrink over the following years; the scenario shown in figure 11.8 assumes a 5 % decline for 1991 (the first year of the deregulation), followed by a further 2 % decline until 93 and a moderate increase of 2 % per annum for the last two years 1994 and 1995.

[6]Indeed one could justify a similar approach for an OECD country, e.g. to quantify the impact from an energy tax. The reason is that it requires a very strong self confidence to believe that the new study (using more or less the same methodology than the previous ones) will result in the true (or even less ambitious, significantly different) elasticities than the thousands of already existing studies.

Deregulation does not only affect conservation but stimulates supply, in particular of crude oil. In fact, the USSR oil industry was highly profitable on any accounting measure but stripped for cash and thus in a miserable shape. Thus deregulation at the wellhead and accompanying support of the international oil companies could mitigate this problem of the ailing oil industry. The technological know how of the oil companies may help to discover new fields from the conjectured but unknown resource basis – the long term effect – and also may increase output in the medium term by introducing improved pressure techniques and secondary and tertiary recovery. The modification of this supply scenario vis à vis the command economy was already shown in figure 11.5.

Figure 11.8 depicts a rosy future: Energy demand declines and even more important, energy intensities are lowered; oil exports can be maintained and gas exports may increase which could provide an amount of the much needed foreign currencies.

Figure 11.8:
Radical Deregulation – Total Primary Energy Requirements and Net Exports of Hydrocarbons

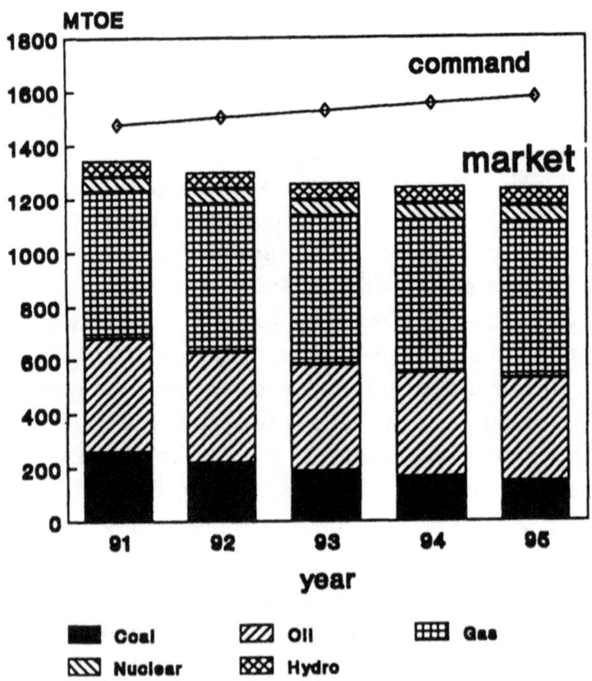

Figure 11.8: (Continued)

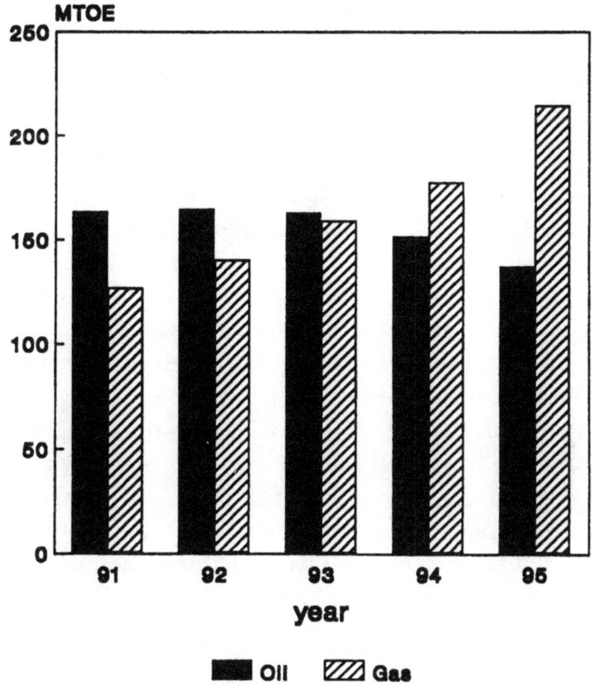

5 Politico-Economical Constraints

The Eastern European countries embark, at least partially, with great enthusiasm on democratic governments and free market economies. However, democracy due to its reliance on majority voting is by no means a guaranty for a free market economy. Majority voting tends to favor the interests of a broad middle class, often for the price of public interventions, regulations or even nationalizations. Furthermore, rent seeking activities of industries, and other lobbies, pork barrel politics, explain the protection that certain, often small, interest groups receive[7].

However, economies at the verge of bankruptcy with no viable economic sector left to finance subsidies provides no scope for rent seeking and could slip into a constitutional phase where efficient rules could emerge along the proposal of Buchanan-Tullock [1962]. Maybe Poland at some points in time

[7]All these effects are subject well documented in the Public Choice literature, see e.g. Mueller [1989].

was close to such a situation when Solidarnosc aimed for a national consensus. However, the tremendous resource endowment of the Soviet Union may attract and provide the feedstock for certain lobbies and pressure groups, which may prove detrimental to an overall reorganization of the economy. In other words, the 'wealth' of the Soviet Union, potential or factual, could provide a serious obstacle to a successful transition.

Therefore, it seems likely that interest groups will spring up that try to maintain some of their old privileges and to seek protection along the transition from a socialistic (and closed) to an open and capitalistic society. The first candidate and definitely largest (in terms of numbers) but weakly organized group consists of the consumers who are accustomed to 'low' energy prices. Now market prices mean real energy price increases for consumers far above the level we observed during the two OPEC oil price shock 1973/74 and 1979/80. Any policy that tries to erode this privilege – radical or gradual – risks political turmoil. Moreover, firms will ask for hidden protection and subsidies through special energy tariffs. Thus, at least the prices for electricity and district heat where the international opportunity costs are not that obvious and the final price for coal, but for different reasons see below, will most likely remain below the assumptions in Section 4.

The deregulation scenario presented in section 4, figure 11.8 shows the diminishing importance of the coal industry, which is far below its potential. The consequence from this decline of demand is that miners have to be fired; productivity improvements further aggravate the necessity to lay off miners. However, miners – contrary to consumers - are usually well organized and mostly highly successful lobby groups around the world. Thus subsidies on domestic coal seem likely indirectly such that demand increases and directly such that (extra-marginal) mines can stay open. However, not only the ailing industries will argue for subsidies but also highly viable industries will engage in activities in order to secure some of the rents for themselves. The prime group I have in mind here are the workers that are employed in the oil and gas industry, which earns a large fraction of the USSR's foreign exchange; there is only one other commodity that yields a comparable amount of foreign currencies, arms exports, which, however, is also in deep trouble. Assuming that these employees, workers and engineers, hold some monopoly power, due to their training and past experience, so that it is costly to lay them off. Then they will (or do they already?) try to capture at least some parts from this foreign exchange through strikes, ignorance and occasional sabotage.

However, the above criticism of democracies is by no means a critique of this kind of a political institution over other political arrangements. On the contrary, facts and theoretical reasoning, most prominent Becker [1983], suggests that democracies with open markets for pressure groups favor 'efficient' measures and are ultimately less populistic than dictatorships (compare also The Economist). This is evident for Eastern Europe, but also for numerous developing countries, where even tough dictators find themselves unable to remove price subsidies of energy and food. In fact, in Latin America democratic governments are able to carry out austerity programs which all the tough 'generals' failed to do.

6 Summary

This paper considered the emerging energy market of the Soviet Union. The first claim is that the continuation of the command economies would lead to an energy demand pattern that would either require oil imports or severe rationing. However, oil imports cannot be financed due to the lack of hard currencies. As a consequence, planners have to turn to the second option, i.e., to ration liquid fuels even more severely. This in turn hampers industrial output due to a lack of energy inputs (and the associated input inefficiency of the command economy) and less motivation of workers (private mobility becomes even more luxurious and privileged). Thus, the planned economy would face severe constraints and potential turmoil, despite the vast energy resources of the USSR. On the other hand, a radical introduction of market prices triggers an enormous conservation potential, facilitates investment in the ailing USSR oil industry (and in the long run tapping the huge oil resources in Siberia) such that the USSR could maintain by and large its traditional oil exports and increases its gas exports. This conservation is associated with a move towards cleaner fuels (predominantly natural gas) such that the present environmental strain is relaxed. However, large blocks lie down the road to a successful deregulation. The correction of the enormous distortion of all prices, not only energy, will be a difficult political task and an impossible task indeed for a phased program. The impossibility of a phased program are manifold, e.g.:

- the expectation of continuous price increases over some time create hoardings;
- a gradual deregulation corrects some deficiencies, thus relaxes the strain to proceed towards a final deregulation;
- pressure groups have enough time to oppose particular price increases.

Ironically, the fact that the Soviet Union is a 'rich' country in terms of resources – oil, gas, coal, uranium, gold, etc. – provides a considerable scope to satisfy aggressive pressure groups and to delay the ultimate deregulation indefinitely.

APPENDIX

A Model of the Energy Demand/Supply in the Soviet Union

We apply a standard framework to model the response of emerging energy markets in Eastern Europe. Table 11.A1 summarizes the model. A unitary income elasticity in the short and in the long run describes the industrial energy demand, such that energy intensities depend on autonomous improvements θ and the price elasticity β. Of course, these improvements proceed sluggishly. A similar but more conventional (except for the non-linearity) dynamic demand relation with constant long run elasticities α for income and β for energy prices – describes the demand within the other two sectors.

Aggregate final, sectoral energy demand is split into secondary fuels through the following procedure. District heat is exempt from fuel competition, because of its characteristic of low marginal costs (due to co-generation) but large and almost entirely sunk costs. The remaining demand is split among solids, liquids, gas and electricity using the multinominal-logit model using price elasticities and an adjustment lag to cater for the sluggish process of fuel substitution. Table 11.A2 outlines the model on fuel substitution; a similar approach applies to power plants' input choices among fossil fuels. The log linear specification (11.A10) suggests the interpretation of β_i as the long run unconstrained share own price elasticity. This is not strictly true but holds only for small shares. Nevertheless, this interpretation of β_i eases the calibration.

Table 11.A1:
Model Equations

SUPPLY:	
Q_t	command economy supply of primary in year t; this is done for each primary energy carrier,
Q^*	(potential) equilibrium supply
S_t	actual supply,
B_t	"boost factor" due to supply-side deregulation,
D_t	Demand (domestic) energy supplier, e.g. coal mines face, domestically plus some exports,
π_t	producer's price at wellhead, mine mouth, etc.,
π_0	base price of the command economy,
σ	long run price elasticity for boost factor,
$(1-\theta)$	fraction of adjustment within a period.

$$Q_t = B_t Q_t^* \tag{11.A1}$$

$$B_t = (1-\theta)(\pi_t/\pi_0)^\sigma + \theta B_{t-1} \tag{11.A2}$$

$$S_t \quad min\{Q_t, D_t\} \tag{11.A3}$$

DEMAND:	
E^*	"equilibrium" or target energy demand,
E_t	(sectoral) energy demand in year t,
y_t	economic activity, period t,
$\varepsilon_t := E_t/y_t$	energy intensity,
ε^*	"equilibrium" intensity, stationary solution of ε_t,
p_t	aggregate price of energy, Laspeyre index,
A	constant,
τ	efficiency relative to the command economy,
α	long run income elasticity,
β	long run price elasticity,
$(1-\delta)$	adjustment within a period,

$$\varepsilon_t^* = A\tau p_t^\beta \tag{11.A4}$$

$$\varepsilon_t = (1-\delta)\varepsilon_t^* + \delta\varepsilon_{t-1} \qquad \text{industry} \tag{11.A5}$$

$$E_t = \varepsilon_t y_t \tag{11.A6}$$

$$E_t^* = A\tau y_t^\alpha p_t^\beta \qquad \text{others, transport} \tag{11.A7}$$

$$E_t = (1-\delta)E_t^* + \delta E_{t-1}. \tag{11.A8}$$

Table 11.A2:
Fuel Substitution

S_t^i	share of fuel i, $i = 1, ..., 4$,
p_t^i	price of fuel i in period t,
a_i	constants,
β_i	'long run share own price elasticity',
μ	adjustment parameter

$$S_t^i = exp(f_i) \sum_{i=1}^{4} exp(f_i) \tag{11.A9}$$

$$f_i = (1 - \mu)[a_i + \beta_i ln p_t^i] + \mu ln S_{t-1}^i. \tag{11.A10}$$

Table 11.A3:
Parameters

Supply	coal	oil	gas	nucl.	hydro
σ		0.10			
θ		0.90			
Aggregate Energy	τ	α	β	δ	
power plants	0.90			0.90	
industry	0.80	1.00	-0.10	0.90	
others	0.90	1.00	-0.10	0.90	
transport	1.50	1.00	-0.10	0.90	
Fuel- Substitution	coal β_1	oil β_2	gas β_3	elect. β_4	Lag μ
power plants	-4.00	-2.00	-1.00		0.95
industry	-2.00	-1.00	-0.50	-0.40	0.95
others	-2.00	-1.00	-0.50	-0.40	0.95

The parameter assessment in table 11.A3 is very cautious, i.e. actual conservation may be significantly larger. First, the price independent impact towards conservation (from tightened budget constraints as discussed above) amounts to a mere 10 % – 20 % improvement, despite we know of the tremendous energy (but in fact overall) inefficiency of the planned economies. Second, price elasticities are far below corresponding estimates for Western economies so that further conservation efforts respond only very moderately to price increases; -0.5 is an educated guess according to EMF [1981]. As a consequence, actual conservation falls still short of potential conservation even for a dramatic increase in real energy prices. Furthermore, adjustment is fairly slow, the time constant of adjustment equals ten to twenty years. The transport sector differs from the other sectors with a negative conservation ($\tau > 1$) due to the fact that private transportation is severely rationed at the beginning so that the demand will increase with deregulation (of course, with the usual ceteris paribus proviso).

The calibration of the fuel share equations is similar to the aggregate energy demand. First, the elasticity parameters constitute 'educated' guesses. These parameters stand for the following hypotheses: (i) solid fuels are most price sensitive because so lids have no particular advantages (except maybe for Iron & Steel); (ii) gas and electricity are much less price responsive, gas for environmental reasons and to accommodate for the the vast gas resources of the USSR, and electricity because it is hardly substitutable for many applications like lighting; (iii) liquid fuels are in between; (iv) power plants are more flexible in their choice of fossil fuels.

References

Buchanan, James M. and Gordon Tullock (1962): "The Calculus of Consent, Logical Foundations of Constitutional Democracy." University of Michigan.

Desprairies, Pierre (1978): "World Wide Petroleum Supply Limits, in World Energy Resources 1985-2020 Oil and Gas Resources." Published for WEC by IPC Science and Technology Press.

EMF#4 Working Group (1981): "Aggregate Elasticity of Energy Demand." *The Energy Journal*, **2/2**: 37–75.

King, Hubbert M. (1978): "World Energy Resources, in Wisconsin Seminar: Natural Resource Policies in Economic Development and International Cooperation." University of Wisconsin, Madison.

Levy, David M. (1990): "The Bias in Centrally Planned Prices." *Public Choice*, **67**: 213–226,

Mueller, D. C. (1989): "Public Choice II." Cambridge, Cambridge University Press.

Nordhaus W. D. (1990): "Soviet Economic Reform: The Longest Road." *Brookings Paper on Economic Activity*, **1**: 287–308.

PlanEcon (1989): "Western Investor's Guide to Eastern Europe and Soviet Union." *PlanEcon Report*, **Vol. 5**.

PlanEcon, Perestroika "Plus": The Albakin Reforms, by Hewett E. A., December 1989.

Popper, Karl R. (1963): "The History of Our Time: An Optimist's View." In: Karl R. Popper, Conjectures and Refutations, London: Routledge and Kegan Paul, pp. 364–376.

Sinyak Y. (1990): "Energy Projections for Eastern Europe and the USSR (1985-2000)." mimeo, International Institute for Applied Systems Analysis, Laxenburg, Austria.

Stern J. (1987): "Soviet Oil and Export to the West."

The Economist, June 29th, 1991, pp. 17–20, Yes, they do march together, but sometimes out of step.

Wirl F. (1991a): "Phasing of Deregulation: Normative versus Positive Objectives." *Journal of Regulatory Economics*, **3**: 89–106.

Wirl F. (1991b): "Impact of the Political and Economical Restructuring in Eastern Europe on the Availability of Net Energy Exports – An Empirical Framework." forthcoming Economics of Planning.

List of Contributors

Wojciech W. CHAREMZA
Economics Department
University of Leicester
University Road
Leicester LE1 7RH
UNITED KINGDOM

John S. CHIPMAN
Department of Economics
College of Liberal Arts
University of Minnesota
1035 Management and Economics
271 19th Avenue South
MN 55455 Minneapolis
U.S.A.

Harald HRUSCHKA
Department of Marketing
Faculty of Economics
University of Regensburg
Universitätsstraße 31
D-93053 Regensburg
GERMANY

Albert JAEGER
FAD/FC Department
International Monetary Fund
Washington DC 20431
U.S.A.

Andrei I. KAZMIN
Commission for Study of Productive
Forces and Natural Resources (KEPS)
Academy of Sciences
Maronovsky per 26
117994 Moscow
RUSSIA

Robert M. KUNST
Institut for Advanced Studies
Department of Economics
Stumpergasse 56
A-1060 Vienna
AUSTRIA

Beverly LAPHAM
Department of Economics
Queen's University
Room A522 MCYN
Kingston
ON K7L 3N6
CANADA

Nina LAPINSKA-SOBCZAK
Department of Economics
University of Lodz
41 Rewolucji 1905 r.St.
PL-90214 Lodz
POLAND

Edward E. LEAMER
The John E. Anderson Graduate
School of Management at UCLA
Department of Economics
405 Hilgard Avenue
CA 90024-1481 Los Angeles
U.S.A.

Judit NEMENYI
Money and Capital Market Dept.
Ministry of Finance
P.O. Box 481
H-1369 Budapest
HUNGARY

George NORMAN
Economics Department
University of Leicester
University Road
Leicester LE1 7RH
UNITED KINGDOM

Baldev RAJ
Wilfrid Laurier University
School of Business and Economics
Waterloo, Ontario
N2L 3C5
CANADA

Thomas URL
Austrian Institute of
Economic Research
P.O. Box 91
A-1103 Vienna
AUSTRIA

Jan VEVERKA
Institute for Forecasting
Opletalova 19
110 00 Praque
CZECH REPUBLIC

Franz WIRL
Institute of Energy Economics
Technical University of Vienna
Gußhausstr. 27-29
A-1040 Vienna
AUSTRIA

Andreas WÖRGÖTTER
Department of Economics
Institute for Advanced Studies
Stumpergasse 56
A-1060 Vienna
AUSTRIA

GPSR Compliance

The European Union's (EU) General Product Safety Regulation (GPSR) is a set of rules that requires consumer products to be safe and our obligations to ensure this.

If you have any concerns about our products, you can contact us on ProductSafety@springernature.com

In case Publisher is established outside the EU, the EU authorized representative is:

Springer Nature Customer Service Center GmbH
Europaplatz 3
69115 Heidelberg, Germany

FSC
www.fsc.org
MIX
Papier | Fördert gute Waldnutzung
FSC® C083411

Kolibri 360 GmbH
Ferdinand-Jühlke-Straße 7
99095 Erfurt, Deutschland
produktsicherheit@kolibri360.de